P9-DXO-501

ALL ABOUT VEGETABLES

Created and
designed by the
editorial staff
of ORTHO Books

Written by
Walter L. Doty

Edited by
Ken Burke

Graphic design by
John Williams
Barbara Ziller

Front cover
photography by
Fred Lyon

Illustrations by
Ron Hildebrand

Ortho Books

Publisher
Robert L. Iacopi

Editorial Director
Min S. Yee

Managing Editors
Anne Coolman
Michael D. Smith

System Manager
Mark Zielinski

Senior Editor
Sally W. Smith

Editors
Jim Beley
Diane Snow
Deni Stein

System Assistant
William F. Yusavage

Production Manager
Laurie Sheldon

Photographers
Laurie A. Black
Michael D. McKinley

Photo Editors
Anne Dickson-Pederson
Pam Peirce

Production Editor
Alice E. Mace

Production Assistant
Darcie S. Furlan

National Sales Manager
Garry P. Wellman

Operations/Distribution
William T. Pletcher

Operations Assistant
Donna M. White

Administrative Assistant
Georgiann Wright

Address all inquiries to
Ortho Books
Chevron Chemical Company
Consumer Products Division
Box 5047
San Ramon, CA 94583

Copyright © 1980
Chevron Chemical Company
All rights reserved under
international and Pan-American
Copyright Conventions.

14 15
—————
89 90

ISBN 0-917102-90-8

Library of Congress Catalog Card
Number 80-66344

Chevron Chemical Company
6001 Bollinger Canyon Road, San Ramon, CA 94583

Acknowledgments:

Copy Editing by
Editcetera
Berkeley, CA

Photography by
Michael Landis
Clyde Childress
William Aplin
Paul Thomas

Additional illustrations by
Ed Bills

Typography by
Terry Robinson & Co.
San Francisco, CA

Color Separations by
Color Tech. Corp.,
Redwood City, CA

No portion of this book may
be reproduced without written
permission from the publisher.

We are not responsible for unso-
licited manuscripts, photographs,
or illustrations.

Every effort has been made at the
time of publication to guarantee the
accuracy of the names and addresses of information sources and suppliers and in the technical data con-
tained. However, the reader should
check for his own assurance and
must be responsible for selection and
use of suppliers and supplies, plant
materials and chemical products.

The author is especially indebted
to Michael MacCaskey for his
assistance in research and
writing the book.

We also thank the following
people for their contributions
to both the first edition and
this edition.

James R. Baggette
Oregon State University

Albert A. Banadyga
North Carolina State University

Maggie Baylis
San Francisco, CA

Russell Beatty
University of California

Louis Berninger
University of Wisconsin

John M. Bridgman
Yorkville, California and
Ripton, Vermont

Gerald F. Burke
W. Atlee Burpee Co.
Riverside, California

Jack Chandler
St. Helena, California

Andrew A. Duncan
University of Minnesota

Eldridge Freeborn
Atlanta, Georgia

James T. Garett
Mississippi State University

A. E. Griffiths
University of Rhode Island
(retired)

Anton S. Horn
University of Idaho (retired)

N. S. Mansour
Oregon State University

John Matthias
San Rafael, California

Charles A. McClurg
University of Maryland

Fred Peterson
Soil and Plant Laboratory
Santa Clara, California

Victor Pinckney, Jr.
Fallbrook, California

Bernard L. Pollack
Rutgers State University
New Jersey

Kenneth Relyea
Farmer Seed & Nursery
Faribault, Minnesota

R. R. Rothenberger
University of Missouri

Raymond Sheldrake
Cornell University (retired)
New York

W. L. Sims
University of California

Perry M. Smith
Auburn University, Alabama

William Titus
Nassau County, New York

Doris Tuinstra
Grand Rapids, Michigan

James Waltrip
Gurney Seed & Nursery Co.
Yankton, South Dakota

Frits Went
Desert Research Institute
University of Nevada

Charles Wilson
Harris Seeds
Rochester, New York

Special food features by
Lou Seibert Pappas
Louise Burton
Annette C. Fabri

ALL ABOUT VEGETABLES

INTRODUCTION

Whether you're growing vegetables in containers on a balcony or planning a garden the size of a football field, here are some helpful hints for finding the information you need as quickly as possible.

Growing vegetables doesn't really have to be complicated. Without ever reading a book (even this one), you could easily go down to your local garden center and buy a bunch of seeds, some fertilizer, some manure, and a bale or two of peat moss. Then, choosing the sunniest spot in your garden, you could spade up the soil, cover it with peat moss and manure, and add fertilizer (according to the directions on the package). After working the manure, peat, and fertilizer into the soil with a cultivator, rake, or rented tiller, you could then begin to plant—simply by following the directions on the various seed packages for planting, thinning, and spacing rows.

Billions of pounds of beautiful vegetables have been grown in just this way by millions of gardeners, using a marriage of trial-and-error and common sense. These two qualities can go a long, long way, and we have no intention of eradicating or complicating them. Indeed, the ideas and suggestions we offer in this book should complement your basic horse sense by helping you avoid some pitfalls, obtain some good, practical knowledge, and spice up your gardening experience.

Just the Facts
If you want to skip the background and get right down to the basic how-to-plant information, turn to the planting chart on page 48 and 49. This chart will not only help you *plant* your garden but also help you *plan* it.

This unlikely but beautiful combination of stock and cabbage shows that vegetable gardening needn't always be formal, but leaves plenty of room for imagination and experiment.

Use the chart for information on *how* to plant as well as on *when* to plant (at least it will keep you from planting peas and beans on the same day; check the columns headed "Needs cool soil," "Tolerates cool soil," and "Needs warm soil"). For more specific advice on planting dates and special treatments, check the major vegetable listings on pages 50-93.

Knowing the length of your growing season, as well as other information about your climate, can prove extremely valuable. To find out, turn to the charts on pages 28-33, locating your town or city (or the one nearest you).

How to Avoid Disappointments
Most gardeners we know have succeeded more than they have failed. When we asked "Did anything surprise you?" one gardener wrote back, "How well things grow under the right conditions and how poorly under the wrong conditions." This answer may hold the key to gardening success. Some of the gardeners we've worked with have shared their successes and disappointments (see "The Spoilers," page 16), as well as their discoveries (see "Good Ideas from Good Gardeners," pages 106-108). Also note the heading "Beginners' Mistakes" as you read about the individual vegetables.

'Rhubarb chard' adds beautiful color to any garden. The leaves are more tender, milder than other varieties of chard.

The daffodils and pansies in this garden tell us it's spring, and the leaf lettuce doesn't seem to be bothered by the competition.

Planning the Garden
There's no one way to do a vegetable garden. The choice of what and where to plant is a highly personal one, reflecting the interests, knowledge, and imagination of the gardener. You may want your garden to be purely practical, or beautiful, or a mixture of both. Some gardeners plant only the most reliable, "success guaranteed" performers, carefully laying out their plots to maximize production. Others remember the unstructured beauty of "grandma's garden" with fascination, and base their plans on that recollection. And some gardeners are full of surprises, always changing their gardens as their own whims change.

"Vegetable gardening" is not quite the same as "growing vegetables." "Vegetable gardening" implies straight rows and an orderly sequence of operations: planning for space, choosing varieties, figuring planting dates, and anticipating harvests, among others. It can be a challenging, even awesome, exercise, especially for the first-timer.

"Growing vegetables," on the other hand, means that you can squeeze your plants into any free corner, set them among flowers and ornamentals, or just plant them for their own special kind of beauty. Imagine a staggered border of blue lobelia and 'Tiny Tim' alyssum; behind that, a row of cabbage with blocks of 'Thumbelina' zinnias between the heads, or a row of marigolds between rows of potatoes for a good contrast near harvest time. 'Salad Bowl' and 'Ruby' lettuce are beautiful with Iceland poppies. Or you might consider red chard with alyssum and 'King Alfred' daffodils. At our California test garden, a favorite combination is yellow violas with parsley. Mixed and planted about 10 inches apart, they make a beautiful sight in spring, with the yellow violas shooting up through the green mountain of parsley. And these are only a few of the many possibilities the vegetable garden can yield.

Whether you're interested in "vegetable gardening" or "growing vegetables," and whether your garden site is as small as a box or as big as a house, pages 35-39 will help you plan your garden.

Interpreting Advice
Pages 50-93 tell you how to grow and care for each vegetable, including: its place in gardening; its particular requirements; special tips; and some of its possible uses after harvest.

Directions are spelled out both here and in the planting charts. For example, take the subject of spacing between plants and rows. Especially if you are a first-time gardener anxious to produce a bumper crop, you may think that the distances presented in the charts are too large and can be cut down. Well, you *can* cheat, but you'll have to pay in the long run—weeding and harvesting become difficult when space is narrow, and since you've robbed the roots of space you'll have to compensate by using more fertilizer and water. Another disadvantage of crowding is that it cuts down air circulation and invites disease.

You don't have to follow our instructions to the letter; there are no perfect recipes for growing vegetables. However, we do try to provide guidelines by presenting throughout the text and in the charts precise dates, temperatures, and measurements for planting, applying fertilizers, adding soil amendments, and other operations.

Don't rely totally on recipes; the garden has too many variables and unknown factors. But don't be discouraged, either—many gardeners who deliberately or accidentally deviate from the recipes still grow perfect vegetables—proof that the measurements are not so critical, after all.

However, this doesn't mean that you should discard all gardening advice and measure by instinct. While it's true that many gardeners let the plants' color, leaf size, or rate of growth tell them when to fertilize, then measure the fertilizer out by the handful or according to how it looks scattered on the soil, these methods are risky. If you wait for the plant to tell you when to fertilize, by then it may be too late to do any good. The biggest—and most frustrating—problem with measuring by instinct is that you have no sure way to duplicate exceptional results, should you get them.

Here, the gardener has planned for both color and texture. 'Ruby Ball' cabbage and marigolds seem to reflect the summer sun. The cabbage did not bolt during the hot days of July and August. "Ruby" is the color of the cabbage head.

How important, then, are gardening directions? And how important is instinct? The answer is that both are of value, but mostly when combined with another, crucial factor: experience. Use the instructions and the measurements (dates, pounds, and inches) as reference points—places to begin and to revisit, as needed; but make all the adjustments necessary for your own climate and soil. Experience is what will help you make those adjustments; and you get experience only by actually growing the plants yourself, in your own garden with its own conditions. Once you have gotten your brain working, your hands dirty, and your enthusiasm sparked, you can begin to consider yourself a good gardener.

Special Advice

It's well known that home-grown vegetables are superior to store-bought corn, snap beans, peas, and all vegetables that lose their fresh sunshine quality soon after picking. This is because vegetables that must be picked green to allow for shipping time lack the taste and quality of ripe, home-gardened produce.

But home-grown vegetables are not automatically superior, as our work with home gardeners and our own test gardens have shown. All factors must come together in the right way; and only when the best varieties are planted, given the right amount of water and fertilizers, and harvested at the right time will the taste be something to brag about: "So much better than that store-bought stuff."

In this book you will read about how to improve your soil, select plants, plant seed, water, fertilize, and otherwise care for a plant growing in garden soil. All these procedures are important, and your mastery of them will doubtless ensure a bumper crop—unless you neglect one small but crucial step in the operation: *weeding*.

Once the vegetable seedlings have emerged, it doesn't take many days for weeds to nearly cancel the good effects of the time and effort you spent planning and preparing the garden.

Hand weeding is a good (though laborious) way to reclaim your garden; and, contrary to popular myth, hand weeding can be enjoyable. You can weed while bending, kneeling, squatting, or sitting, which will give you an almost eye-level view of the plants and soil. And when you pull a grassy weed that is robbing a seedling of water and nutrients, or even threatening to strangle it, you are performing an act of kindness that is sure to reward you with healthier plants and a more fruitful harvest.

Working with Nature

Gardening books like to talk about plants that supposedly can't be grown in your climate. For the most part this advice is accurate, but we've met too many determined gardeners who take great pleasure in growing what can't be grown. If your interest in the plant is strong enough, you can grow it—even if you have to create a special climate around it.

On pages 25-27, you'll find information on how to alter unfavorable environments; how to work with nature's whims and tricks; and how to swing with nature's rhythms.

The Rare and Unusual

Each year, more unusual vegetables are made available to seed companies. If you're looking for a hard-to-find plant, you'll find it keyed by number to the list of seed companies on page 104. Items carried by four or more of the companies are not keyed.

A Final Note

As we have suggested, don't blindly accept the recipes and directions in this book or any other. Only the plants in your garden can tell you the truth—and the plant is always right, no matter what any book or authority has said. Dr. Frits Went, a noted scientist and horticulturist, puts it this way:

"Once the amateur has realized that he himself is master of the situation in his garden, and that he is not the slave of a set of recipes, a great deal is gained. Gardening comes out of the realm of mystic beliefs and becomes an adventure in adaptation. Each plant grown becomes an experiment, instead of a routine performance. That plant becomes the test of whether the applied principle was right. If the plant does not grow well or dies, the application of the principles was not right, or the conditions were such that the principle did not work. If, on the other hand, the plant behaves well, it shows the applicability of the principle.

"By looking at the plants in this way, a garden becomes immensely interesting, it becomes the testing ground of ideas, and it frees the mind from dogmatism. The gardener becomes aware of the fact that experiments can be carried out everywhere, and are not restricted to highly specialized laboratories.

"Science is not a cult; it flourishes where these observations are faithfully recorded."

This planting of flowering cabbage produced many colors and forms. These cabbages need near freezing weather to show color and should be planted for late fall-winter maturity.

THE FUNDAMENTALS OF GROWING VEGETABLES

No matter what you're growing, there are fundamental principles that will ensure gardening success: good soil, a continuous and uniform supply of water and nutrients, and protection of leaves and roots. This chapter gives you all the basics you need.

Plants require certain basic ingredients in order to grow: soil that will retain air once the water has drained out; water, a continuous and uniform supply; fertilizer, a continuous and uniform supply; and a chance to grow, free of damaging insects and diseases.

SOIL

If you are one of those lucky gardeners whose soil is deep, fertile, easy to work, and easy to manage, you can skip to "Watering" on page 12 right now—any remarks we make about soil are not for you. However, most home gardeners are not this fortunate; many have problem soils. Some of these soils are too coarse, but most of them are medium- to fine-textured loams or clays. If your soil is fine-textured, don't despair—your vegetables will thrive if you water them frequently and deeply enough to maintain the required moisture. Just be sure not to overwater, so that you don't deplete the air space. In addition, you can add organic matter to make your fine-textured soil more friable. This may not greatly affect the life of the vegetable, but it will make *your* life easier.

Air in the Soil

One of the most important factors determining plant growth is how much air is in the soil. Roots breathe, just as we do, taking in oxygen and breathing out carbon dioxide. If roots (or even portions of roots) are deprived

Plants vary widely in their requirements of the basics: soil, nutrients, and water. Corn, for example, needs more moisture than some, and a good supply especially at maturity. This chapter gives you the general needs of vegetables, the chapter "The Vegetables" the specifics for each.

of air, they will die of suffocation.

In fine-textured garden soils, the space between soil particles (*pore space*) is very small. When water is applied to the soil, it drives out air by filling up the small pore spaces. Then, when the plant wilts from lack of air, the unwary gardener takes pity on it and waters it some more, killing it with kindness.

Roots need air, and whenever they are deprived of it you can expect trouble. There is a surprising variety of ways to deprive roots of air—more than there are ways to fill the pore spaces with water.

When a soil crusts over after a hard rain, it partially cuts off the air supply to the roots. When the top layer of soil is compacted, the soil underneath suffers for lack of its full requirement of air. Thus will a well-worn path across a lawn actually kill the grass.

To get a clearer idea of the effects of compaction, just look at the trees in your community, in school grounds, or in parks—any place where foot traffic is heavy. If the soil is packed down on the surface (compacted), this means that the tree is struggling for survival. Compacted soil not only reduces the supply of air to the roots, but it also reduces the amount of moisture available to the plant, since water runs off the surface soil instead of soaking in.

Boxes designed to show below-ground action help us to understand the big job a seedling has to perform. If the soil crusts over, the emerging plant may be unable to break through; if the structure of the soil is too dense, the first tender roots may have difficulty in growing.

Soil Tests

Soils may be ideal for one kind of plant but less than ideal for another. A soil in which rhododendrons and azaleas will thrive will not support tomatoes, for example. To grow the best vegetables you can, it's worthwhile having your soil tested.

One important thing you will find out from a soil test is the pH level of your soil. This, quite simply, is a measurement of acidity and alkalinity. The values run from 0 to 14, with 7 representing neutrality. Numbers less than 7 indicate increasing acidity, and numbers greater than 7 indicate increasing alkalinity.

For optimum development, most vegetables require a pH level of about 6.0 to 7.5. (Potatoes are one exception —they prefer a slightly acid soil.) If your soil is too acid, the soil test lab will recommend lime; if your soil is too alkaline, the lab will suggest soil sulfur.

If lawns in your area receive yearly liming treatments, you can be sure that your own soil, too, is naturally lime-deficient. This is true for gardeners living east of Mississippi in the South, or east of Ohio in the North. In western areas where annual rainfall is low, alkaline soils tend to be the general rule.

Many states offer free or low-cost soil tests. Take advantage of this good investment and call your local County Extension Agent for details. (As of this writing, only California and Illinois lack a soil-testing service. If you live in either of these states, contact one of the many private laboratories.)

Improving Your Soil

Amending less-than-perfect soil takes some effort, but not an enormous amount. You don't need to redo the entire garden area—it's enough to add organic amendments to only those areas in which you intend to plant.

When you mix organic matter with fine-textured soil, the soil becomes more mellow and easier to work. Organic matter—for example, compost, peat moss, manure, sawdust, or ground bark—provides three benefits: it opens up fine-textured soils; it improves drainage in the amended portion; and it allows air to move more readily through the soil, thus warming it up earlier in the spring. In lighter, more coarse-textured, sandy soils, organic matter performs the function of holding moisture and nutrients in the root zone. Sandy soil has a limitless capacity for organic matter—the more you add, the more you increase the soil's moisture-

Almost any problem soil can be improved by adding organic material. This can be anything from compost to manure, and generally, the more you add the more you improve the soil.

and its nutrient-holding ability.

How much organic matter should you add? Enough to change the physical structure of the soil. This means that at least a third of the final mix should be organic matter. If you spread at least a 2-inch-thick layer of organic matter over the soil and work it in to a depth of 6 inches, you will have the desired result.

There is no point in adding just "a little dab" of anything—miniscule amounts won't change the soil structure. A little peat moss or straw and compressed clay soil will make only an adobe brick. So use enough organic matter. If it's peat moss you intend to use, be sure to moisten it first with warm water or a "wetting agent" (a soap-like material designed for this purpose). Then mix it into the soil.

Synthetic Soils

If you'll be growing vegetables in containers, your best bet is a quality-controlled, synthetic soil mix. In our test gardens years ago, we filled raised beds with these lightweight mixes. After many vegetable seasons have gone by, we still find much to say in the mixes' favor.

You can buy special soil mixes at garden stores everywhere. They go

by such trade names as: Redi-Earth, Jiffy Mix, Metro Mix, Super Soil, Pro-Mix, Baccto Potting Soil, and Terra-Lite Tomato Soil. Some mail-order seed companies package their own brands of potting and special soils. For example, Burpee Seed Company offers its own special tomato mix, Burpee Tomato Growing Formula; and Park Seed Company makes its Sure-Fire Mixes. Other seed companies may list either their own mixes or widely distributed commercial mixes.

Ingredients: The organic part of the mix may be peat moss, redwood sawdust, shavings, bark of hardwoods, fir bark, pine bark, or a combination of any two or more.

The mineral part may be vermiculite, perlite, pumice, fine sand, or a combination of two or more. The most commonly used minerals are vermiculite, perlite, and fine sand.

Vermiculite (Terra-Lite) resembles mica, when mined. When it is heat-treated, its mineral flakes expand with air spaces up to 20 times their original thickness.

When perlite (Spong Rok is one brand) is mined, it is a granite-like volcanic material. When this is crushed and heat-treated, it pops like popcorn and expands to 20 times its original volume.

The mix you buy may be 50 percent peat moss and 50 percent vermiculite; 50 percent wood products and 50 percent fine sand; or some other combination of organic and mineral components. But although the ingredients may vary, the principle behind all mixes remains the same—soilless "soil" must provide:

—Fast drainage of water through the soil.

—Air in the soil after drainage.

—A reservoir of water in the soil after drainage.

The most important element in any mix intended for containers is how much air is left in the soil after drainage. Container soil must drain better than garden soil. In the garden, water moves through a column of soil with continuous capillary action (like a blotter). But in a container, that continuity is broken by the existence of the container bottom, gravel, or whatever impedes the water's downward flow; and where the continuity is broken, the water builds up.

Artificial soil mixes circumvent this problem. When the mix is watered, the water drains quickly from the macropores (large pores), thus allowing air to follow the water down to the roots. At the same time, however, some of the water is retained in the micropores (small pores) of the mix.

The quality soil mixes are ready to use as is. Just fill the container with the mix, set the transplant in it, add enough water, and watch it grow.

However, feel free to tamper with your store-bought mix, if you choose. You can add sand (10 percent by volume) to weight down the mix, or you can use perlite to lighten a mix that tends to be too fine.

Keep it simple: Some gardeners like to work out complicated mixes containing five or six ingredients, but this isn't necessary. A simple combination of peat moss and perlite or fine sand gives good results.

If you're planning to use many containers or fill a few raised beds, it's more practical to make your own mix. Here is a recipe for 1 cubic yard (27 cubic feet) of such a mix.

  9 cubic feet of fine sand
18 cubic feet of ground bark or nitrogen-stabilized sawdust
             or
  9 cubic feet of fine sand
  9 cubic feet of peat moss
  9 cubic feet of ground bark
To either of the above formulas, add:
  5 pounds of 5-10-10 fertilizer
  7 pounds of finely ground dolomitic limestone
  1 pound of iron sulphate

Compost

The black, fragrant, crumbly, partially digested organic residue called compost comes from garden waste material. Whatever method of composting you use, the main objective is to arrange the waste material in such a way that the soil organisms that break the waste down can thrive and multiply. These organisms need moisture, air circulation, and food.

To build the pile, use a mixture of green and dry materials. Green material decomposes rapidly—grass clippings, lettuce leaves, pea vines, and other succulent materials contain sugar and proteins that provide excellent nutrients for the organisms. On the other hand, dry material—sawdust, dry leaves, small twigs, and prunings—contains very little nitrogen and decomposes very slowly when composted alone. By mixing green and dry materials, your pile will compost at just the right rate.

The size of the woody material will affect the rate of decomposition. If dry leaves and other dry materials are put through a shredder rather than added to the pile as is, the small pieces will decompose faster (more surfaces are exposed to the decay organisms). Shredding also creates a fluffier mixture, making air and water penetration more efficient. You can buy or rent a shredder, or you can just use a rotary mower to shred leaves. If you produce grass clippings in large quantities, mix them thoroughly into the composting material. Other-

Vermiculite

Sand

Bark

Perlite

Peat Moss

Jiffy-Mix

Top: This three-part compost bin has proved to be an excellent design in one of our test gardens.
Above: Medium-sized redwood bark covers the walkways between mini raised beds. The black plastic mulch hastens maturity.

wise they will begin to form a soggy mass, putrefy, produce unpleasant odors, and attract flies. After you've mixed the grass into the compost, spread a layer of soil or old compost over the top.

For the sake of convenience, divide the compost pile into three piles. The first is for the daily collection of waste products—refuse from the vegetable harvest, wastes from the kitchen, coffee grounds, egg shells, small prunings, and so on. The second is for the fast-working compost; add nothing to this pile, and make sure to turn it frequently. The third pile is for the finished compost.

Mulches

Mulches and organic amendments may be made of the same materials; how they differ has to do with when and where the application is made. A mulch covers the soil on top—it is not worked in—and is applied only after the soil has warmed fully in spring. Mulches offer a variety of benefits: They protect roots in the top inches of soil from high temperatures; they conserve water by reducing evaporation from the soil; and they prevent erosion and reduce soil compaction caused by foot traffic or by water from heavy rains or sprinklers. And a summer mulch, tilled back into the soil in the fall after harvest (or the next spring before planting), will improve the soil. (For more information on mulches, see page 42.)

Raised Beds

The raised bed is one of the oldest ideas throughout gardening history. Anywhere soil and water problems have made good gardens difficult

the solution has involved planting above ground-level. The variations are many, but the results are almost invariably good.

A basic advantage of raised beds is that you can choose which kind of soil to fill the bed with. To avoid stressing the plants for water or nutrients, you can build a raised bed and fill it with a loose, fast-draining soil mix (see page 11).

If your area typically receives heavy spring or fall rains, you should definitely consider raised beds. These, unlike ground-level gardens, will permit vegetable growing. Since water drains quickly in a raised bed, the roots won't languish in waterlogged soil.

It's easy to use fertilizer in a raised bed. You know exactly how much area is to be covered. For instance, if you have two beds, each 4 feet wide and 12 feet long, the two areas total 96 square feet. When the fertilizer instructions are stated in pounds per 100 square feet, it's easy to be on target.

WATER

Few gardening subjects stir up as much confusion and controversy as this familiar liquid. How much water should be applied? When? How often? If your tomato blossoms drop, you may be told it's because of "too much water." But someone else may tell you it's because of "too little water." If your carrots come out stumpy, you may be warned about overwatering. If your lettuce tastes bitter, you may be cautioned against giving an uneven supply of water.

Since every gardener has a different answer, how can you tell what's too much and what's too little?

Overwatering

"Too much water" can mean either (1) too much water for the roots to grow in, or (2) too much water to produce fruit (producing, instead, maximum leaf growth).

Obviously, you must water your plants, but you also must take care not to overwater them. Plant roots need both moisture and air in order to grow. Therefore, they require a growing medium that will allow for the penetration of air, which will bring oxygen to the plant and remove carbon dioxide from it. Overwatering will fill all the air spaces in the soil with water, thus stopping root growth. The longer the soil is deprived of air, the greater the damage will be. Once damaged, roots are prey to rot-causing soil organisms, which can cause the plant to die of root rot. Is it any wonder that gardening experts often call water "the hazardous necessity"?

To truly be kind to your plants, water just often enough to accommodate your soil's water-holding capacity.

Soils on the clay side have a high water-holding capacity; the air spaces are very small, allowing water to move through them slowly. Such soils need to be watered only infrequently. Observing and feeling the soil beneath the surface for several inches will help you gauge when to water. Of course, the best way to deal with the watering issue is to prepare your soil properly. (See "Soil," page 9.)

Underwatering

On the other hand, if you apply "too little water," according to irrigation specialists your plant will be under *water stress*—a malady that progresses from slight to severe. A thirsty plant must work harder as the moisture supply in the soil decreases.

Many plants have a remarkable ability to recover from water stress. When impatiens, for example, suffers from lack of moisture, it drops its flowers, droops down, and appears to be dead. However, one lavish application of water will be sufficient to make it straighten up within only a few minutes. Soon afterwards, it will begin to develop a new crop of flowers.

Occasionally, gardeners use water stress as a deliberate tactic—for example, when growing herbs for a seed crop. And many flower-producing annuals respond to water stress by flowering even more profusely. To the annual plant, a dry spell is a signal to get busy with the business of flowering and seed-setting before it's too late.

However, water stress rarely pays off in the vegetable garden, except in unusual cases—for instance, if a tomato plant is producing only vines, it will grow and ripen fruits when put under water stress. But for the most part, putting vegetables under stress will exact a severe penalty. Snap beans will drop their blossoms. Lettuce—whose shallow root system requires a steady supply of moisture—will turn bitter. Cucumbers will stop growing altogether (although they will resume if watered again). Beets will get tough and stringy. Radishes will turn hot. Turnips will develop too strong a flavor. Muskmelons will lose their sweetness.

Flowers may fully recover from the retarded growth caused by water stress, but vegetables will not. For a successful harvest, make sure that these vegetables receive water during the following critical periods:

An oscillating sprinkler set in the middle of a garden satisfies many gardeners. Others complain about muddy walkways.

Vegetable	Critical Period
Asparagus	Brush—Fern
Broccoli	Head development
Cabbage	Head development
Carrot	Root enlargement
Cauliflower	Head development
Corn	Silking and tasseling ear development
Cucumber	Flowering and fruit development
Eggplant	Flowering and fruit development
Lettuce	Head development
Lima bean	Pollination and pod development
Melon	Flowering and fruit development
Onion, dry	Bulb enlargment

Proper Watering

You already know about the hazards of overwatering and underwatering, but what about just plain watering properly? Again, there are no exact formulas, but the basic rule of thumb is to water thoroughly, filling the root zone. Then let the soil dry out a bit, and when it looks like it needs it, water again.

How can you tell what "thorough" watering means to your specific garden? Apply what you think is a sufficient amount of water. Then, using a spade or shovel, dig up the top 3 or 4 inches of soil and take a look. Did the water penetrate this far down? If yes, you are watering properly. If no, you need to water longer.

You can get a feel for how long to wait between waterings by observing the appearance of your soil, and using your hands to feel it.

As to when during the day to water, any time will do. However, many experts suggest the early morning, which offers two advantages: (1) the plants will lose less water to evaporation, and (2) the plants will stay dry at night, and thus be less susceptible to attack by disease-causing organisms.

When watering plants in containers, keep applying the water until it drains out the bottom, which will leach nitrogen and potassium through the soil mix. This may seem like a waste of fertilizer, but it offers a distinct advantage by preventing harmful salts from building up in the mix.

Furrow Irrigation

This is probably the most commonly used method of watering vegetables. Furrow irrigation is frequently preferred by gardeners who have a variety of vegetables crowded into a small area. Overhead sprinklers cannot be selective enough to avoid those

Deep furrow irrigation confines the water to the roots.

Drip emitters are good for new plantings.

Viaflow soaker-oozer permits uniform watering of the root zones.

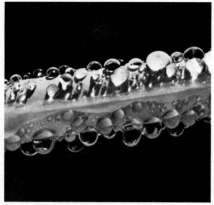
Close-up view of the viaflow system, showing how water passes through the small pores.

vegetables that do not like wet leaves.

Furrow irrigation confines the water to the plant's root zone, and also inhibits weed growth in unwatered areas

When you lay out your furrow irrigation system, keep the furrows as level as possible. Too much slope will cause the water to flow too rapidly, and some areas may receive too little water. You can, if you have the room, place the furrows to allow dry walkways between paths.

Drip Watering
The idea of making a little water go a long way is nothing new, especially in areas with a limited supply of water. What *is* new, however, are some irrigation techniques and equipment. Emitters, spot-spitters, Dew-Hose, Jumbo-oozers, Viaflow, Twin-wall, and Drip-Eze are a few of the names a modern gardener may hear.

The drip/trickle system of irrigation is now available to the home gardener, after years of testing in thousands of acres of orchards, row crops, and nursery operations. Although these systems are not totally foolproof, their potential advantages are so great that it seems worthwhile to experiment with them in the home garden.

Dr. Falih Aljibury, irrigation and water technologist at the University of California, describes drip watering in this way:

"While generally considered as a new irrigation method, the basic concept of drip irrigation has been practiced since the beginning of this century by nurseries growing fruit trees and ornamental plants. However, it has only been in the past few years that this concept has been expanded to include application in many crops grown in the field as well as under nursery and greenhouse conditions.

"With drip irrigation, the water drops onto the soil surface without disturbing the soil structure, so that the water can sweep between soil particles. Once in the soil, the water moves by capillary action to the surrounding areas.

"Drip irrigation drops the water onto the ground through one or more emitters located adjacent to each tree or plant.

"In drip irrigation one tries to replenish the water on an almost daily basis, this amount of water being equal to the water used by the plant since the last irrigation. In other words, drip irrigation does not store water for a long future-plant use, but rather constantly replaces water that has already been used."

Sprinklers
There are two kinds of sprinklers—underground sprinkler systems and hose-end sprinklers. The underground systems are good for large gardens, as they provide even coverage with a minimum of guesswork. If installed on a timer, they will water with convenience and regularity. For smaller gardens, you can stay with underground systems or go to hose-end sprinklers. The latter will cost less, although they offer less convenience. They come in many sizes and shapes. Choose the one best suited to your area, the one that will use water most efficiently.

Fertilizers
On a daily basis, plants require only a small amount of nutrients, but that amount must be available just when the plant needs it. Some slow-growing plants allow some leeway—you don't need to fertilize until you see the lower leaves begin to yellow. But vegetables won't let you be this casual. Right away, they demand adequate nutrients in the soil to see them through to harvest. An insufficient amount of nutrients will retard growth, and this, in turn, will reduce both quality and yield.

How Much?
When you apply fertilizers, whether fish emulsion, blood meal, commercial liquid, or commercial dry fertilizers, be sure to follow the label directions on the package or bag. Don't try to outguess the manufacturer. If you must err, err on the side of "too little." Too much of any fertilizer, even manure, is dangerous.

All commercial fertilizers are labeled by the percentages they contain of nitrogen, phosphorus (as P_2O_5), and potassium (as K_2O). There are many formulas—5-10-5, 5-10-10, 6-18-6, 16-16-16, and so on—but the percentage of nitrogen is always indicated by the first number. The nitrogen is the most important element—the percentage of it in the formula dictates how much fertilizer to apply. The phosphorous and potassium just tag along.

If the fertilizer you are using contains one of the higher percentages of nitrogen, use less of it. Read the directions on the package to find out the nitrogen percentage, or else you may unintentionally double or triple the amount needed.

The following chart shows how the amount of fertilizer to be applied decreases as the percentage of nitrogen increases. (Assume that 5-10-10 fertil-

Some gardeners enjoy hand
watering and have the time for it.

leaching action of the water drains away part of the nutrient supply. An experienced container gardener will feed vegetables with a weak nutrient solution continuously, taking a cue from the watering schedule rather than from the calendar.

Ways to Apply Dry Fertilizer

Before planting, mix the fertilizer with the soil. Spread it evenly over the soil at the rate called for on the bag or box, and work it in with a spade or tiller.

Apply the fertilizer in narrow bands in furrows 2 to 3 inches from the seed and 1 to 2 inches deeper than you will be placing the seeds or plants.

Apply the fertilizer as a side dressing, after the plants are up and growing. Scatter it on both sides of the row 6 to 8 inches from the plants. Rake it into the soil and water thoroughly. Banding is one way to satisfy the need of many plants, especially tomatoes, for a steady supply of phosphorus as the first roots develop. When fertilizers are broadcast and worked into the soil to a shallow depth, not all of the phosphorus is immediately available to the plant. By concentrating the phosphorus in the band, you give the plant what it needs.

Tips to Remember

—Be sparing with all types of fertilizers. Too much manure can cause as much trouble as any other kind of fertilizer. When using manure, cut down on the rate of fertilizers.

—Follow label directions with liquid fertilizers.

—Follow up dry fertilizer applications with a good watering to dissolve the fertilizer and carry it into the root zone.

—When using large amounts of fertilizer for such heavy feeders as cabbage and onions, apply half the amount before planting, and then side dress with the remainder once growth is underway.

—Check the timing for applying fertilizer for each vegetable. The first application is crucial with many crops, since it can protect early vigorous leaf growth.

—Give less nitrogen to plants grown in partial shade than to the same kind of plants grown in full sun.

—Increase the amount of fertilizer when plants are crowded by narrow spacing between rows, and when plants are grown in a random pattern in a small plot.

izer recommendations call for 3 to 4 pounds per 100 square feet.)

Formula	Pounds per 100 Square feet
5-10-10	3.5
6-20-10	2.8
8-24-8	2
10-10-8	1.7
16-16-16	1

When applying dry fertilizer, it's easy to figure the amount to apply per square foot or length of row if you use liquid measurements. A pint of dry fertilizer weighs about 1 pound, 1 cup weighs ½ pound, and so forth.

Container plants: Vegetables need a small but continuous supply of nutrients. When the weather is hot, you end up watering frequently; then the

Applying Dry Fertilizers

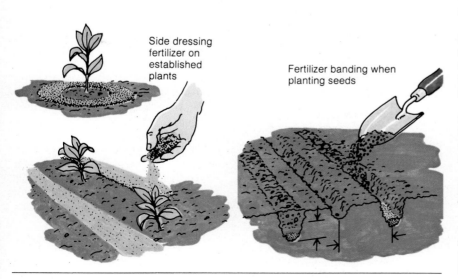

Side dressing fertilizer on established plants

Fertilizer banding when planting seeds

Portable wood and wire frames offer protection from birds, rabbits, squirrels, and other garden pests. When covered with a plastic film, as shown here, they also serve as row greenhouses.

THE SPOILERS

In an attempt to find out what difficulties vegetable gardeners tend to meet up with, we asked people throughout the country the following questions:

—"What disappointments did you have in last year's garden? Crop failure? Mistakes? Why?—or where do you think you went wrong?"

—"What success did you have?"

—"Did anything surprise you? We are interested in both good and bad surprises—as well as what caused them, if you know."

The gardeners who replied covered a wide range. Some had given a great deal of attention to insecticides and fungicides. Others had tried to ignore insects and diseases totally. The following answers are typical of the many we received. *Note that not all the spoilers were insects or diseases.*

From New Jersey

Failure: "I watered and weeded and felt we would have good vegetables but we were disappointed. Worms in the radishes. Tomatoes not so good."

What varieties did you buy? "Don't remember, went mostly by the pictures."

From Massachusetts

Failure: "We had a great deal of rain and cold weather during May. Quite a few different kinds of seeds

didn't come up and had to be replanted. The bugs ate the plants as fast as they came up."

Success: "The cucumbers and pole beans and corn are going good now after the second planting."

From New York

Failure: "Trouble with fusarium wilt on tomatoes, and corn borer and corn earworm."

Success: "Insect control other than corn. I tried to stick to a 10- to 14-day schedule through the growing season."

Did anything surprise you? "Results from succession of plantings; replanting of areas already planted."

"The amount of produce you can obtain from a 10- x 20-foot garden."

"How much better my garden looked than the organic garden next door."

From Oregon

Failure: "Radishes were very poor. Didn't use enough Diazinon in treating soil."

Success: "All other crops were great. Found that late carrots do beautiful and you can leave them in the ground most of the winter."

Failure: "Pepper plants were planted in too much shade. A combination of too much water and too rich a soil from compost. I may have made this year's tomatoes grow too much foliage —not enough producing fruit."

Success: "Peas, this year and last, grew very well. I planted in mid-February and they grew fast despite snow and frost. I grew and reported on seven types of experimental peas for the county agent."

Did anything surprise you? "In the small shaded area I use, I am often surprised at the amount the area can produce. I grow a little bit of everything there."

From Louisiana

Failure: "None to speak of."

Success: "Produced 27 vegetables —over $900 (at retail) worth on a 45- x 60-foot garden."

Did anything surprise you? "I marvel at nature's way of producing vegetables with a little help from the gardener."

From Northern California

Failure: "In some areas of my garden (particularly shady areas), root maggots completely destroyed crops of radishes and turnips, though they never seemed to harm carrots. What's the cure? Tomato wilt killed one plant last year and is threatening almost half my plants this year. I have heard that some varieties are more resistant to the disease than others. Is that true ... and which ones?"

Success: "'Country Gentleman' corn grew over 12 feet high and made for very good eating (we will plant a shorter variety of corn this year so that it doesn't cut out sun for the rest of the garden). Our bush 'Buttercup' plant is a *giant* (bush plus runners)! Bush beans were so heavy they needed staking."

From Southern California

Failure: "Summer crop was pretty much a failure. Tomatoes suffered from wilt and failure to set fruit. Green peppers and eggplant not good. Bush beans started out well then turned yellow and died. Failure attributed to too much horse manure,

too much water, too many insects, too many gophers."

Success: "The results with zucchini and Swiss chard were excellent. The winter crop of beets, cabbage, broccoli, radishes, lettuce, and turnips were excellent."

From Florida

Failure: "Poor okra germination. The consensus was that soil temperatures were too low for best okra seed germination."

Success: "Generally good."

What should beginners do? "(1) Have their soil pH checked early (add dolomite lime if needed); (2) Plant varieties suggested by State Extension Service; (3) Water properly; (4) Don't overfertilize; (5) Use correct pesticides."

PESTS AND DISEASES

Following are ways to control the most common pests. To find out about additional insects and diseases, check with your County Agricultural Agent for information on problems special to your area, or send for the 50-page booklet, *Insects and Diseases of Vegetables in the Home Garden* (Superintendent of Documents, Government Printing Office, Washington, D.C. 20402).

When using insecticides on vegetables, check the label to find the number of days before harvest to stop spraying. This number differs according to the kind of vegetable. If your garden is small and planted with different, mixed-in vegetables, you can't provide each kind with a separate treatment. In that case, stop spraying or dusting two weeks before harvest.

Aphids

Although aphids feed on many vegetables, they are most damaging to members of the cabbage family—broccoli, cauliflower, Brussels sprouts, and cabbage. Growth in damaged plants is stunted, often distorted. Different-colored aphids denote different species; they appear as grey, black, brown, or red.

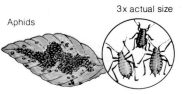

Aphids — 3x actual size

Control: Try simply washing off the aphids with a water spray. If that isn't sufficiently discouraging, use Malathion or Diazinon products labeled for vegetable garden use.

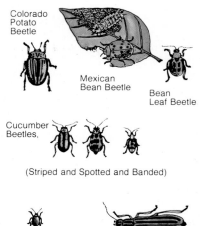

Colorado Potato Beetle · Mexican Bean Beetle · Bean Leaf Beetle

Cucumber Beetles, (Striped and Spotted and Banded)

Flea Beetle · Blister Beetle · actual size

Beetles

This incredibly diverse group of insects includes harmless, damaging, and beneficial types. Ladybugs are among the beneficial variety, since they eat garden pests; the cucumber beetle *(Diabrotica)* is one of the most troublesome. Other pestiferous beetles are illustrated.

Control: Many chemical controls are available, but take care to use only those that are made specifically for vegetable gardens. Products containing Sevin and Methoxychlor frequently are recommended.

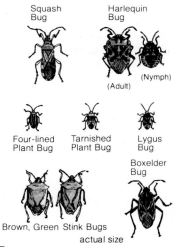

Squash Bug · Harlequin Bug (Adult) (Nymph)

Four-lined Plant Bug · Tarnished Plant Bug · Lygus Bug

Boxelder Bug

Brown, Green Stink Bugs · actual size

Bugs

Nongardeners may consider any insect a "bug," but when gardeners talk about bugs, they mean a specific group of insects, many of which are significant garden pests. Stink bugs, for example, attack carrots, lettuce, okra, and peppers. Squash bugs damage squash and pumpkins.

Control: The best control is with dusts containing Sevin, Methoxychlor, or Rotenone. Apply while the bugs are still young; mature bugs are considerably tougher to kill.

Caterpillars and Worms

These are the larvae of various moths and butterflies. They come in all sizes and colors, some hairy and some with spines. All have healthy appetites and are among the most frequently damaging vegetable-garden pests. The cabbage looper and tomato hornworm fall into this category.

Control: Dusts containing Sevin or Methoxychlor are effective and convenient, or use the biological control, *Bacillus thuringiensis*. Note all label cautions and directions.

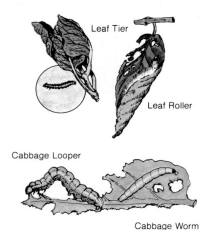

Leaf Tier

Leaf Roller

Cabbage Looper

Cabbage Worm

actual size

Corn Earworm and Tomato Fruitworm

Both these names refer to the same insect; which one gets used depends on which crop is attacked. The corn earworm causes the most damage to sweet corn in the United States. The worst damage occurs during the silk stage, when the eggs hatch on the silk and the larvae begin eating their way into the developing ear, destroying the kernels and leaves. The tomato fruitworm attacks the flowers and developing fruit of the tomatoes.

Control: Since corn earworms overwinter in the soil as a pupa, thorough soil cultivation in late fall will aid control. Look for sweet corn varieties with some resistance to this pest. Feeding worms can be stopped with insecticide dusts that contain Sevin.

Tomato Fruitworm

Corn Earworm
⅔ actual size

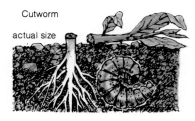

Cutworm

actual size

Cutworms

The cutworm, a type of caterpillar, is extremely prevalent in the garden. During the day, it lives just beneath the soil surface; at night, it emerges to feed on tomatoes, cabbage, peppers, beans, and corn, cutting them off at the soil line.

Control: Thorough cultivation in late summer or fall will expose and destroy many cutworm eggs and pupae. You also can make a collar out of an empty 12-ounce tin can, sticking it about an inch into the soil to protect the young plants. Or use a granular or dust product containing Diazinon before planting.

Leafhoppers

Leafhoppers are small (⅛ to ½ inch), wedge-shaped insects that feed by piercing and sucking. Although they are general feeders, they particularly damage beans, lettuce, potatoes, squash, and tomatoes. They also will feed on bean blossoms, causing a poor pod set. Often the most severe leafhopper damage is caused not by their feeding, but by the virus diseases they spread.

Control: Reflective mulches are helpful, especially when plants are young. Sprays containing Pyrethrins, Rotenone, or Sevin also are effective.

Leafhoppers   2x actual size

Leafminers

Leafminers

These troublesome insect larvae live inside plant leaves and mine the plant tissue, frequently rendering the leaves useless. This reduces the plant's vigor and, of course, the harvest. Peppers, tomatoes, cucumbers, melons, and squash often are attacked by these pests.

Control: Use a dust or spray containing Diazinon, according to label directions and precautions.

Mites

Mites are a kind of tiny spider whose webbing on the undersides of leaves often is more visible than the insects themselves. Thriving in hot, dry, and dusty conditions, mites make a speckling on the top of damaged leaves. This always reduces the plant's vitality; sometimes it destroys the plant altogether.

Control: Use a Malathion spray on beans, peas, broccoli, and Brussels sprouts. If melons and squash are attacked, use a Diazinon spray.

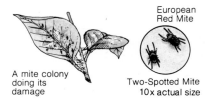

European Red Mite

A mite colony doing its damage

Two-Spotted Mite
10x actual size

Onion and Radish Maggots

These are the larvae of flies that appear in spring and lay eggs on the soil near the base of vegetables. Onions and radishes are the two favorite targets, although other vegetables are attacked, as well.

Control: A fine-mesh wire cloth over the seedling row will prevent the adult flies from laying eggs. Or use a granule or dust product containing Diazinon. Control for the cabbage root maggot is similar.

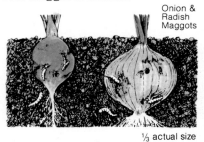

Onion & Radish Maggots

⅓ actual size

Slugs and Snails

These are among the most ubiquitous and damaging vegetable-garden pests. They feed mostly at night, hiding in cool, damp locations during the day. The silvery slime trail they leave behind is the mark of their activity.

Control: Search them out at night by flashlight and pick them by hand; use traps; search their hiding places during the day; or use baits. The safest baits for vegetable-garden use are those containing Metaldehyde.

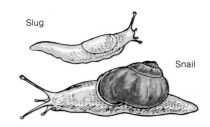

Slug

Snail

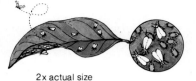

White Flies

2x actual size

White Flies

Adult white flies are small (1/16-inch long), wedge-shaped, and pure white. They fly like clouds of snowflakes when disturbed. "Nymphs," which are even smaller, do the most damage. Scale-like, flat, and oval, they appear as pale green, brown, or black. They stunt plant growth by sucking the juices from the leaf undersides.

Control: Use Diazinon or Malathion products intended for use on food crops. Follow label directions and be sure to spray the leaf undersides, where most white flies hide.

Preventing Diseases

Planting disease-resistant varieties is the surest way to prevent diseases from devastating your crop. Since more disease-resistant varieties are bred each year, you have a wide choice.

If you have already had—or want to avoid—trouble with diseases, check the variety list of vegetables in this book (we have noted those that have some disease resistance) and in catalogs. Be especially alert for mention of resistance to the diseases that damage these vegetables:

Tomatoes—Fusarium, verticillium.
Cucumbers—Scab, mosaic, downy mildew, powdery mildew, anthracnose.
Muskmelon—Fusarium, powdery mildew.
Snap beans—Mosaic, powdery mildew, root rot.
Cabbage—Virus yellows.
Spinach—Blight, blue mold, downy mildew, mosaic.

Cornell's Raymond Sheldrake, our consultant, emphasizes the importance of resistant varieties:

"So many people tell me that their cucumber vines just up and die, even though they sprayed them. Well, the big problem is that their plants became infected with mosaic, a virus disease known as cucumber mosaic virus. The only way to escape it is to use mosaic-resistant varieties.

"In my radio programs I stress, for example, mosaic resistance in cucumbers and suggest mosaic-resistant varieties, but when someone who heard the program would go to a garden store, all they might find is a pretty picture of a cucumber on the outside of a seed packet. The variety might be 'Straight 8' with no resistance to anything, but the picture on the front of the package is what sells it. It's an open-pollinated cheap seed. Education is a slow process and I think we are closing the gap between what our breeders are doing and what our gardeners are using, mainly because some of these diseases have become so severe that the gardener just can't get by with any old variety."

Rotation. Controlling diseases by rotating crops is good, standard gardening advice. The most effective method is to change the garden site every few years. Second to that is rotating the crops within the garden so that the same crop doesn't occupy the same space year after year. (See the rotation plan in "Good Ideas from Good Gardeners," page 106.) However, it's difficult to rotate crops in a 10- by 20-foot garden.

The Good Guys

These are some of the "good guys"—the beneficial insects that prey on aphids, mites, caterpillars, bugs, and other harmful insects. If you should find them in your garden, welcome them and protect them as the friends they are.

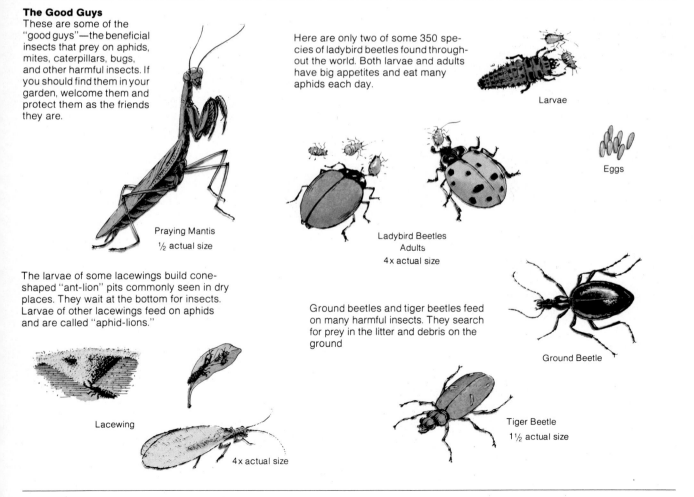

Here are only two of some 350 species of ladybird beetles found throughout the world. Both larvae and adults have big appetites and eat many aphids each day.

Larvae

Eggs

Praying Mantis
1/2 actual size

Ladybird Beetles
Adults
4x actual size

The larvae of some lacewings build cone-shaped "ant-lion" pits commonly seen in dry places. They wait at the bottom for insects. Larvae of other lacewings feed on aphids and are called "aphid-lions."

Ground beetles and tiger beetles feed on many harmful insects. They search for prey in the litter and debris on the ground

Ground Beetle

Lacewing

4x actual size

Tiger Beetle
1 1/2 actual size

YOUR CLIMATE AND THE VEGETABLE GARDEN

Special regionalized charts will help you determine the length of your growing season, and pinpoint first and last frosts in your area. Discussion of sun, shade, cloud covers, fog, rain and wind will help you utilize your climate to your best gardening advantage.

Vegetable climates are measured with a different set of yardsticks than those used to measure the climates for vines, shrubs, and trees. For shrubs, an important factor in climate adaptation is minimum winter temperatures. But with vegetables, sub-zero winter temperatures don't matter as long as there are enough days between the last frost of spring and the first frost of fall.

YOUR VEGETABLE GARDEN CLIMATE

Pages 28-33 show the growing seasons of various US cities. To make use of this chart, look for your city (or the city nearest you) to get an idea of the growing-season duration in your garden.

Just how your own garden will vary from the information on these charts depends on many climatic factors, including the following.

Season Length

Season length is one of the most important factors determining what and how much you can grow. The growing-season lengths in the US and Canada vary all the way from barely 3 months to as much as 12 months. They are indicated by the colored bands on the charts, pages 28-33.

Day Length

The length of the day or number of hours of actual sunlight influences the growth habits of several annual vegetables, particularly spinach and Chinese cabbage. The lengthening

Although it makes a pretty picture, frost is not always a welcome sight for vegetable gardeners.

of days beyond the 12-hour day/night cycle of the vernal equinox signals these vegetables to flower. Temperatures begin to rise, making the environment more favorable. Sometimes, however, vegetables will "bolt" if the day length isn't just right—that is, they will go to seed before they are ready to harvest.

It's best to plant when the day length is most favorable for your particular vegetable. When Chinese cabbage is planted during the long days of spring, it rarely grows properly. But when it's planted to mature in the short days of fall, it is as easy to grow as any other vegetable.

Spinach, on the other hand, requires short days; long days cause bolting, especially if the plants were subjected to cool conditions when small. To avoid this, plant in early spring, and use a "long-standing" variety. Or, in mild-climate areas, plant in the fall.

Lettuce also tends to bolt. This can be caused by the lengthening of days, but more often it has to do with the increasing warmth of the temperature.

Breeders of vegetables have applied considerable skill to the problem of bolting; and to avoid or reduce the losses caused by bolting, they have created "bolt-resistant" varieties. Look for them when you are shopping for vegetables that might otherwise tend to bolt.

Vegetable gardeners need to be aware of the number of clear, sunny days they can expect in their growing season. Cloud cover and fog can affect the performance of plants.

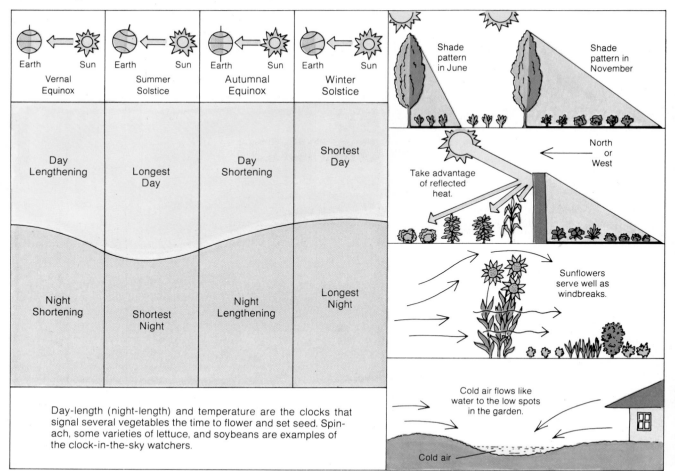

Day-length (night-length) and temperature are the clocks that signal several vegetables the time to flower and set seed. Spinach, some varieties of lettuce, and soybeans are examples of the clock-in-the-sky watchers.

Shade and Fog

If you have a tree or two and want to grow vegetables, you may be spooked by the advice given to gardeners since the first garden book was written: "Locate the vegetable garden where it will receive full exposure to the sun."

Full sun is, of course, ideal, but gardeners have grown many kinds of leafy vegetables quite successfully in partial shade. The USDA House and Garden Bulletin No. 163, *Mini-Gardens for Vegetables*, states that the following vegetables can tolerate partial shade: beets, cabbage, carrots, chives, kale, leeks, lettuce, mustard, green onions, parsley, radishes, Swiss chard, and turnips. However, all these belong to the cool-season group. Warm-season growers—such as peppers, eggplant, and melons—require full sunlight.

One of our consultants, A.E. Griffith of the University of Rhode Island, suggests that the words "partial shade" should be qualified. Partial shade is one thing in a climate with consistently clear, sunny days, he observes, but quite another in an area subjected to frequent fog or cloud cover. In New England, for example,

Mr. Griffith has noted the following:

—"Most vegetables do best only in full sun in the relatively cool, stormy summer climate of New England."

—"Most vegetables will tolerate shade for two or three hours per day, although they will not perform quite as well as they should."

—"Much of the coast of New England from eastern Connecticut eastward to New Brunswick is subjected to fog or cloud overcast in May and June, as well as in September and October. At best, vegetables receive only about 65 percent of the total sunshine that's theoretically possible throughout the growing season."

—"Low solar energy plus partial shading will often seriously delay the maturation of many warm-weather crops."

The cloud cover in western Washington creates a similar situation. For example, tomato trials in Mt. Vernon revealed a most interesting picture of seasonal variations due to a long, cool summer. When several tomato varieties in the Puget Sound area were tested for "days to maturity," it was found that not one variety had ma-

tured at the advertised rate. An early variety such as 'Early Girl' may require as little as 45 days between transplanting and the first ripe fruit. However it required 98 days—more than twice as many—in the Mt. Vernon test. The other tomato varieties tested produced these results in terms of typical maturation rates and the number of days required at Mt. Vernon.

Variety	Average Number of Days	Mt. Vernon Number of Days
Early Girl	45	98
Jet Fire	60	119
New Yorker	64	112
Springset	65	112
Fantastic	70	119
Jet Star	72	119

These figures demonstrate how local growing conditions can influence vegetables' growth, and why you have to take all vegetable-growing recipes (no matter how authoritative they may appear) with a grain of salt. By making certain adjustments or modifications in your gardening regimen (many of which are discussed here), you can grow most vegetables anywhere.

Radiation Principles

The sun is the source of the earth's climate. Radiant heat in sunlight may be either good or bad, depending on your geographic location, the season, and the air temperature surrounding you.

As solar radiation (composed of light and heat) reaches the earth, a number of things happen to it. Some is reflected into space from clouds. Some is scattered and diffused in the sky as it strikes the dust and water vapor in the air. Small amounts are absorbed by carbon dioxide, water vapor, and ozone in the earth's atmosphere. All this accounts for 80 percent of the sun's energy that reaches the earth. The remaining 20 percent reaches the surface of the earth, where it is either absorbed or reflected.

As a result, the earth may receive solar radiation as direct radiation from the sun, as reflected radiation from atmospheric particles in the sky, or as reflected radiation from materials on the surface.

Absorption, radiation, and reflection: At night the ground continues to radiate heat absorbed during the day, thus cooling the ground. The absorption qualities of various materials can be used to increase heat storage or decrease it, as desired. For instance, the orchardist floods the ground to reduce frost potential; the water produces heat, which slows outgoing radiation. (However, use this technique with caution: evaporative cooling from sprinkled leaves on damp soil can chill a surface.)

The gardener's favorite insulation material, however, is mulch. Loose mulches, such as grass clippings, leaves, or straw, reduce the radiation that reaches the soil.

Rocky areas, pavements, and masonry also can make extremely hot pockets, thereby radiating a little sun onto cold areas nearby.

Color also has an effect on radiation —a light-colored wall will bounce more light to nearby plants than a dark-colored wall will.

Slope of the Land

Cold air flows like water. As the air next to the ground cools at night, it becomes heavier. Like water, it flows slowly downhill into washes, valleys, or cold-air basins, filling depressions and low spots as it progresses.

In a western canyon, it's not unusual to see pines, oaks, and other temperate-zone trees growing on the northern slope, while the southern exposure has only a covering of grass or some sparse desert vegetation. The

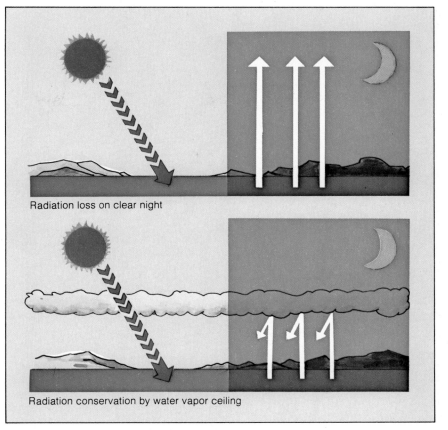

Radiation loss on clear night

Radiation conservation by water vapor ceiling

When the sun goes down, insolation stops and heat loss from the ground begins. Radiation is much faster to a cold high ceiling than a low warm one, and day-night temperatures under them can vary as much as 40 degrees.

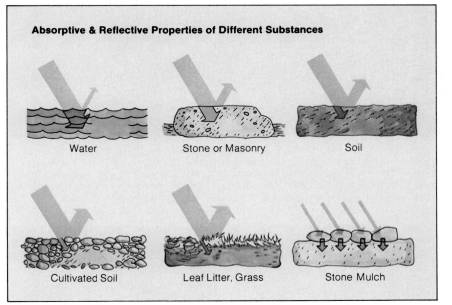

Absorptive & Reflective Properties of Different Substances

Water

Stone or Masonry

Soil

Cultivated Soil

Leaf Litter, Grass

Stone Mulch

Water is by far the best heat storage reservoir we have. The ocean absorbs 95% of the insolation it receives, reflecting little back into the atmosphere. It is a very slow radiator, and its day and night temperatures are more constant than those of faster radiating land surfaces. The density of granite, and other stone, and masonry, holds their absorption capabilities at higher levels ranging from about 50% to 70%. Average soil absorbs and stores about 30% of the insolation. When cultivated, however, the figure drops to about 20%, as the air spaces created are poor conductors of heat. The capacity of light soil is less than that of dark soil, as light colors reflect more and absorb less than dark ones. The capacities of sand and peat are greater when damp, with their air spaces filled with water. Grass and leaf litter, because of their many air spaces, are at the bottom end of the scale with only 5% storage capacity. Snow is very poor, because of its air content and highly reflective qualities.

The problem: In the only space available for a planting of corn an 8-foot hedge stopped the morning sun; a 10-foot hedge put the garden in shade in late afternoon. In the first experiment aluminum foil was stapled to plastic and hung on both east and west sides of planted area. Corn crop was perfect except in rows next to hedge. The next year reflective panels of metalized plastic with adjustable angle of reflection were more efficient.

Above: Clear plastic can be used as a windbreak as well as for soil mulch or a row cover.
Right: This portable hinged A-frame increases spring soil temperatures. The open end permits air circulation.

air temperatures may be similar on both slopes, but there is a wide variation in ground temperatures, and it is these temperatures that regulate the kind and amount of life on the opposite sides. Northern slopes also retain moisture longer.

Slopes can be miniature banana belts or potential frost zones, depending on their direction of pitch. You can use a lath structure, a controlled area of filtered sun, to moderate uncomfortable winds (frequent problems on mountainsides). Barriers of plants or structures, properly located, also will temper prevailing winds.

Elevation
Temperatures cool as the elevation increases. As a rule of thumb, a 1000-foot increase in elevation is the equivalent of the climate 300 miles north (at the same elevations and slopes).

Nature's Rhythm
You can avoid frustration when growing vegetables if you accept the plant's natural rhythm rather than trying to make it fit yours. Spring fever in the first warm days of the season makes for a delicious, heady sensation, but not for the best guide on when to plant. The best way to enjoy your vegetable garden is to be in step with *its* needs, not yours.

Check the planting chart on pages 48-49 to find out whether the vegetables you mean to plant are in the cool-season or warm-season group. Some of the cool-weather crops may need to be planted before you really feel like gardening.

"Cool season" means more than that the vegetable can be planted early; it has to do with quality. The good taste of peas and beets, for example, depends on the temperatures at which they ripen. The first picking of garden peas and the first harvest of beets are the great ones. When the first warm days of spring roll around, it may seem ridiculous to wait for the soil to warm up before planting beans and corn, or to wait for the night temperatures to rise before setting out tomato transplants. But it's worth the wait—if the soil temperature is below 55°, beans will rot, and tomatoes and eggplant will sit and sulk. Lima bean seed is likely to rot if the soil is below 62°; the same goes for okra.

Climate and Man, the USDA yearbook, lists these ideal temperatures for various vegetable groups:

Distinctly cool-region crops that prefer 60°-65°F. and are intolerant of high-summer temperatures above a monthly mean of about 70°-75°F.:

—Very hardy—Cabbage and relatives, kale, sprouting broccoli, turnips, rutabagas, kohlrabi, spinach, beets, and parsnips.

—Damaged by frost—Cauliflower, heading broccoli, lettuce, carrots, celery, peas, and potatoes.

Crops adapted to a wide range of temperature but not tolerant of freezing:

—Prefer 55°-75°F. and tolerate some frost—Onions, garlic, leeks, and shallots.

—Prefer monthly means of 65°-80°F. and will not tolerate frost—Muskmelons, cucumbers, squash, pumpkins, beans, tomatoes, peppers, and sweet corn.

Distinctly warm-weather, long-season crops that prefer a temperature mean of about 70°F. and tolerate no cool weather—Watermelons, sweet potatoes, eggplant, some peppers, and okra.

—Perennial crops—Asparagus, globe artichoke, and rhubarb.

Extending the Season

A determined gardener can always "cheat the season" by improving the climate, especially for specific crops. You might need a warmer or cooler season, or more or less moisture, or less sun, or less wind. Whatever your need, there are devices that can help you fool Mother Nature.

Raising the Temperature

You can warm up your soil's temperature in several ways. You can grow (or build) a tall, reflective surface—a row of corn or sunflowers, or a tall fence—and plant warm-season crops just south of that. Or you can plant on the south side of a ridge.

To protect your seedlings from the cold, you can use plastic cottage cheese containers or others of that type, or plastic jugs with the bottom removed. Polyethylene plastic film, both clear and black, also will raise the soil's temperature.

Top: A portable A-frame trellis can be moved anywhere in the garden to take the most advantage of the sun or find protection from winds. In addition, you can grow tomatoes, cucumbers, peas, beans, or almost any vining vegetable on the trellis.

Center: The A-frame can be large enough to work in and still be portable, and it can be covered with chicken wire and clear plastic to become a kind of greenhouse.

Bottom: Extension agent Duane Hatch has proven that a clear plastic mulch will add the extra warmth needed to ripen melons in western Oregon's cool summers.

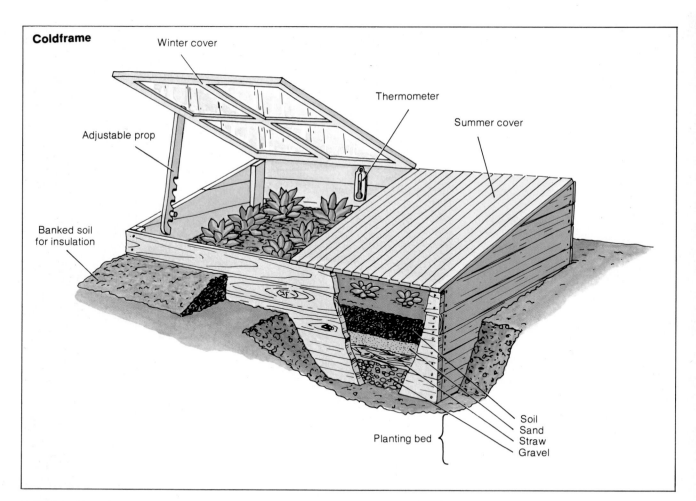

Coldframe

Winter cover

Thermometer

Summer cover

Adjustable prop

Banked soil for insulation

Planting bed

Soil
Sand
Straw
Gravel

The Coldframe

At the turn of the century, no farm garden was complete without a coldframe—a bottomless, usually glass-covered box that's heated only by the sun. This air-tight box, made of redwood or pressurized-treated wood, is sunk into the ground. Its hinged, transparent window can be made of stock coldframe "sash" (available from greenhouse supply firms), an old window or glass door, or fiberglass or polyethylene film.

Orient the coldframe toward the south in order to receive maximum winter sunlight. Painting the inside walls white or silver helps reflect more light to the plants.

As far back as 1897, W. J. Wickson wrote:

"A coldframe is simply for the purpose of concentrating sun heat and protection from low temperatures and heavy rains. It is a convenient receptacle for seed flats, or it may be put over seeds sown in the ground. The frame is normally made of one-inch boards about twelve inches wide. The back board or boards are about 18 inches wide and the sides slope from this 18 inches to 12 inches at the front.

"The frame is usually made 3 feet deep and is then covered with glazed

One of our test garden coldframes has exchangable panels. Fiberglass (top) retains heat. Lath (center) allows air exchange. Shade cloth (bottom) reduces light and wind to protect seedlings.

sash or cloth frames or lath frames or first one and then the other, according to the amount of protection and heat or shade desired. The arrangement is called a 'coldframe' because no provision is made for bottom heat. There are many modifications of the coldframe. Lath or slat houses or lath covers for beds with raised edging boards are all on the coldframe principle. In warm climates where so little increment of heat is required and where shade is often necessary, the modified coldframe is well suited."

Coldframe uses. Many vegetable gardeners use a coldframe to start seeds early. A coldframe lets you start most garden seeds up to eight weeks earlier than in garden soil. In short-season areas, those eight weeks really count.

If your winter is not too severe, you can grow winter salad greens easily in a coldframe. Lettuce, chives, and most other greens will grow through until spring, if protected in this way. The bottom heat that the coldframe provides will carry the greens through the coldest winters.

Coldframes also can be used for forcing bulbs and propagating a variety of flowers, azaleas, and evergreen and deciduous trees and shrubs.

A Climatic Experiment

Using a coldframe in one of our test gardens, we tried to determine the following:

—Would a coldframe let us grow a vegetable crop not well adapted to our western climate (in this case, sweet potatoes)?

—Would a coldframe improve the setting of fruits on tomatoes and peppers?

—How much frost protection would a coldframe provide?

—What is the difference in reradiation between clear and black plastic? (Since clear plastic allows unimpeded radiation to the plant and soil during the day, theoretically, black plastic should slow reradiation to the cold night sky.)

—How many ways could we improve seed germination? (We'd already tried row covers and modified styrofoam picnic boxes.)

With these questions in mind, we set up some experiments. We used soil and maximum/minimum thermometers to keep track of temperatures both inside and outside (1) the "black and white" experiment, (2) the fiberglass coldframe, and (3) the automatic coldframe. (Daily maximums/minimums are as reported locally.) The results are given in the table below.

Night temperatures: Vegetables

The author checks his commercial automatic coldframe. It opens 6 to 7 inches when air temperatures exceed 72°F and closes when they go below 68°F. No electricity is required—the automatic device is a simple thermocouple.

such as tomatoes, eggplant, and peppers require a night temperature no lower than 55°F. to set fruit. They may not get this at the 4:00 A.M. minimum, but they might get it as late as 10:00 P.M. This is why providing small amounts of protection that will slow reradiation will contribute to an earlier harvest.

Here's how 10:00 P.M. temperatures compared with reported minimum temperatures in our garden:

Day	Daily Maximum Temperatures	Daily Minimum	Black & White 10:00 P.M.			Fiberglass Coldframe 10:00 P.M.		Automatic Coldframe 10:00 P.M.	
			Clear	Black	Out	In	Out	In	Out
1	95	49	66	68	60	62	58	64	60
2	86	48	62	63	58	59	55	60	57
3	77	54	59	62	57	60	55	59	57
4	73	56	60	61	58	62	58	61	59
5	70	54	63	64	60	61	58	62	60
6	77	51	65	67	63	64	61	65	63
7	82	49	67	69	63	64	61	65	63
8	89	56	62	64	58	64	60	60	58
9	84	50	65	66	62	65	59	62	60

Day	Nightly Minimum	10:00 P.M. Temperature
1	45	52
2	46	55
3	50	58
4	53	60
5	52	56
6	54	54
7	54	60
8	50	56
9	50	60
10	51	60
11	52	58

VEGETABLE CLIMATES OF THE SOUTH

In the highest elevation and the shortest growing season in the South there may be as few as 160 growing days. But this is more than enough to grow corn and tomatoes as well as a wide variety of both early spring and fall vegetables. You can plant lettuce, endive, turnips, Chinese cabbage, and kale in midsummer for fall and early winter harvest.

In the long-season gardens of much of the South (220 to 250 days), the fall and winter garden can stretch the harvest season. The Oklahoma State University in a fact sheet on fall gardens says: "A small plot, well tilled, will produce fresh vegetables from September 10 to December most years."

Tender vegetables	
Summer squash	August 10-20
Winter squash	July 15-25
Cucumbers	August 10-20
Beans (Bush lima)	August 10-20
Beans (Bush snap)	August 10-20
Beans (Pole)	July 15-25

Semi-hardy vegetables	
Irish potatoes	August 1-15
Carrots	August 1-15
Swiss chard	August 1-15
Beets	August 1-15
Lettuce	August 1-15
Turnips	August 1-15
Spinach	Sept. 10 - Oct. 10
Radishes	Sept. 10 - Oct. 10
Mustard	Sept. 10 - Oct. 10
Winter onions	Sept. 10 - Oct. 10

Growing Season: 220 Days or Less

Inland areas have a fairly high amount of clear days, while the coastal portions see less of the sun. The amount of rainfall varies greatly throughout the climate zone, with the western area receiving 20 or more inches annually while the eastern areas along the coast get twice that amount. The climate is versatile enough to allow spring and summer plantings of cool-season crops. The summer season is long enough and temperatures high enough to allow the commercial growing of tomatoes and cowpeas.

The 220-day growing season here is long enough to permit summer plantings of snap beans, carrots, sweet corn, and tomatoes; as well as fall plantings of the cool-season crops.

Plantings for fall and winter harvest should be in by early August.

Growing Season: 220 to 250 Days

There is considerable variation in climate from the south to the north portion of this zone. Cool-weather crops are planted in mid-February. Plantings of muskmelons and squash are delayed until mid-March. A succession of plantings is possible for bush, snap, and lima beans, both in spring from mid-March through May and again from July 1 through mid-August.

A long growing season of 250 days or more, under the influence of the South Atlantic and Gulf Coasts, gives the garden a definite spring and fall planting for the majority of crops. Broccoli, Brussels sprouts, Chinese cabbage, and kale are planted in late summer—August and September. Warm-season crops, such as muskmelons, corn, and peppers are planted early to avoid the hottest days and nights of summer.

Growing Season: 250 to 365 Days

A preference for late summer and fall plantings is found in the warmest portions of this climate zone. September is the best date for escarole, endive, lettuce, radishes, broccoli, and other cool-season crops. However, winter frost is recognized, and snap beans, muskmelons, sweet corn, and peppers are not planted until the soil warms up in mid-February. Onion varieties planted in this zone are bred for planting in September, to bulb in the short days of late winter and spring.

In the south portion of Florida, the normal planting season of northern states is reversed, starting with the cool-season vegetables in the early spring. "Early spring" is September for broccoli, cabbage, lettuce, English peas, and potatoes. October is "spring planting time" for beets, carrots, radishes, turnips, and spinach. Tomatoes are a winter and summer crop with a planting season from August to March. The starting dates for the warm-weather vegetables such as pole beans, sweet corn, peppers, cucumbers, and squash begin in January.

VEGETABLE CLIMATES OF THE WEST

Season lengths in the West vary all the way from less than 100 days to virtually all year—360 days. But especially in the West, season length does not tell the whole story. For instance, note that Fresno and Seattle have about the same length of growing season. August temperatures in Fresno are above 90°, whereas in Seattle they barely reach 70°.

Just as in the South, this zone does have some high-elevation, short-season gardens. Carrots, beets, turnips, onion sets, and potatoes yield good crops if planted as soon as the ground can be worked. All cool-season leafy vegetables that can be harvested before they are mature can be grown successfully.

Spring weather may fluctuate from sub-freezing at night to the upper 60s or 70s by mid-afternoon. Many areas have only 60 frost-free days a year; some can expect frost every month.

In the 100-day areas, the last frost date in spring is often around May 30. July temperatures average 58° to 66°F.

In these high-elevation areas there's no chance for sweet corn. But winter squash, peppers, and the long-season leafy vegetables that can be harvested before they mature are okay. Most of the root crops (turnips, carrots, beets, etc.) are the best bet.

Growing Season: 160 Days or Less

All the above-listed vegetables are easily grown with this season length. The early varieties of tomatoes, corn, and other warm-season vegetables can be added to the list.

This area includes those gardens east of the Cascades and at the lower elevations in the Rocky Mountains. July temperatures here of 66° to 72° are just right for the maximum growth of a wide variety of vegetables. Clear days, low humidities, and high light intensities promote rapid plant growth. From June through August the area receives more than 80 percent of possible sunshine. In the shortest growing season, the early maturing varieties of the warm-season crops should be selected.

Growing Season: 160 to 200 Days

Areas with season lengths within this range vary widely. There is sunny Pocatello, ID; high Denver, CO; and high desert Albuquerque, NM.

In areas such as the Columbia basin and the Lewiston and Boise valley areas of eastern Washington and Idaho, the growing season is about 190 days, and summer temperatures are warm. With the heat, high light intensities, and low humidities, the gardeners can successfully grow melons—even okra and peanuts.

In Utah, the St. George area enjoys a 200-day growing season, with July temperatures in the 80s.

In the high desert areas of New

Climates of the South

City	July % of Sunshine	Days of Growing Season	Inches of Rain	Days of Rain	Last Frost	First Frost	JAN.	FEB.	MARCH	APRIL	MAY	JUNE	JULY	AUG.	SEPT.	OCT.	NOV.	DEC.	
Roanoke, VA	59	165	25"	66	4-14	10-26	45/25	47/27	54/32	65/40	74/50	81/58	85/63	84/62	77/55	67/43	55/34	47/27	
Asheville, NC	59	195	26"	69	4-12	10-24	49/30	51/31	57/36	68/44	76/52	83/60	85/64	84/63	79/57	69/46	57/36	50/30	220 days or less
Lexington, KY	64	198	24"	66	4-13	10-28	43/25	45/27	55/34	66/43	75/53	84/62	87/66	86/64	81/58	69/47	54/36	45/27	
Lubbock, TX	81	205	15"	40	4-1	11-9	53/26	57/29	65/34	74/44	82/54	91/64	92/66	92/66	84/58	75/47	62/33	55/27	
Richmond, VA	65	218	29"	68	3-29	11-2	48/29	51/29	59/36	70/46	79/55	87/63	89/78	86/65	82/58	71/47	60/37	50/29	
Louisville, KY	73	220	25"	68	4-1	11-7	44/26	46/28	56/35	67/45	77/54	85/63	89/67	88/65	82/58	71/47	55/36	46/28	
Knoxville, TN	64	220	25"	73	3-31	11-6	50/33	53/33	60/39	70/48	79/56	86/65	88/68	87/67	84/61	73/49	59/38	50/33	
Tulsa, OK	76	221	27"	56	3-25	11-1	46/26	52/30	60/36	70/47	78/58	87/67	93/71	93/70	86/62	75/51	59/36	49/30	
Okla. City, OK	78	224	25"	56	3-28	11-7	46/27	51/30	60/37	70/48	77/58	86/67	92/70	93/70	85/62	73/51	59/37	49/30	
Nashville, TN	69	224	24"	68	3-28	11-7	49/31	51/33	59/39	71/48	80/57	88/66	91/70	90/68	85/61	74/49	59/38	51/32	
Texarkana, AR	77	233	28"	64	3-21	11-9	52/32	57/37	64/42	74/52	82/61	90/69	94/72	93/71	87/64	78/52	64/42	56/36	
Ft. Smith, AR	74	234	29"	60	3-21	11-10	50/29	55/33	63/40	74/50	81/59	90/67	95/71	94/70	87/62	77/51	62/38	52/32	
Raleigh, NC	62	237	30"	71	3-24	11-16	52/31	54/32	61/38	72/47	79/56	86/64	88/68	87/67	82/60	73/48	62/38	52/31	From 220 to 250 days
Memphis, TN	73	237	28"	62	3-20	11-12	51/33	59/35	61/42	72/52	81/60	89/69	97/72	92/70	86/63	76/52	62/40	53/35	
Baltimore, MD	64	238	26"	68	3-28	11-19	44/25	45/26	54/32	66/43	76/53	83/61	87/66	85/65	79/58	68/46	56/34	46/26	
El Paso, TX	78	238	6"	33	3-26	11-14	56/30	62/36	69/40	78/49	87/57	95/67	95/69	93/68	88/61	79/50	66/36	58/31	
Birmingham, AL	62	241	33"	75	3-19	11-14	57/36	59/38	67/42	76/50	84/59	91/68	92/71	92/70	89/64	79/52	66/41	58/36	
Little Rock, AR	71	241	31"	66	3-17	11-13	51/31	55/34	63/41	74/51	82/60	90/68	93/71	92/70	86/62	76/50	61/38	52/32	
Atlanta, GA	62	242	29"	72	3-21	11-18	54/36	57/37	63/41	72/50	81/59	87/66	88/69	88/68	83/63	74/52	62/40	53/35	
Norfolk, VA	66	242	34"	79	3-19	11-16	50/32	51/32	57/39	68/48	77/58	85/66	88/70	86/69	81/64	71/53	61/42	52/33	
Dallas, TX	78	244	23"	51	3-18	11-17	56/36	59/39	67/45	75/55	83/63	91/72	94/75	95/75	88/67	79/57	66/44	58/38	
Augusta, GA	62	249	34"	74	3-14	11-1	59/36	61/37	67/42	76/50	84/59	91/67	91/70	91/69	86/64	78/52	68/40	59/35	
Colombia, SC	63	262	34"	74	3-14	11-21	58/36	60/36	66/42	76/51	85/60	91/68	92/71	91/70	86/65	77/52	67/41	58/34	
Shreveport, LA	79	262	29"	67	3-8	11-15	57/38	60/41	67/47	75/55	83/63	91/71	93/73	94/73	88/67	79/55	66/45	59/40	
Wilmington, NC	66	262	40"	79	3-8	11-24	58/37	59/38	65/43	74/51	81/60	87/68	89/71	88/71	84/66	76/55	67/44	59/37	
Houston, TX	70	262	38"	80	3-14	11-21	64/44	66/46	72/51	78/59	86/66	91/72	92/74	93/74	89/69	82/60	71/51	65/46	
Savannah, GA	61	275	41"	89	2-27	11-29	62/41	64/42	70/47	77/54	85/62	90/69	91/71	91/71	86/67	78/56	69/45	63/40	
Montgomery, AL	66	279	38"	77	2-27	12-3	59/38	61/40	67/45	76/52	84/61	90/69	92/72	92/71	88/66	79/55	66/42	59/38	From 250 to 365 days
New Orleans, LA	58	292	47"	93	2-20	12-9	64/45	67/48	71/52	78/58	84/64	90/71	91/73	91/73	87/69	80/61	70/50	65/46	
Charleston, SC	66	294	44"	94	2-19	12-10	61/38	63/40	68/45	77/53	84/62	89/69	89/72	89/71	85/66	77/55	68/44	61/39	
Mobile, AL	61	298	58"	90	2-17	12-12	62/44	65/46	70/50	77/58	86/65	91/72	92/73	91/73	87/68	80/60	70/48	64/44	
Jacksonville, FL	62	313	48"	103	2-16	12-16	67/45	69/47	73/51	80/57	86/65	91/71	92/73	91/73	88/71	80/62	72/51	67/46	
Crps. Christi, TX	80	335	27"	73	1-26	12-27	67/47	70/51	74/56	80/63	86/69	90/74	93/75	94/74	90/71	84/64	74/54	69/49	
Tampa, FL	61	349	51"	105	1-10	12-26	71/51	73/53	76/56	81/61	87/67	89/72	90/73	90/74	89/72	84/66	77/57	73/52	
Miami, FL	65	365	60"	127	—	—	76/58	77/59	80/61	83/66	85/70	88/74	89/75	90/75	88/75	85/71	80/65	77/59	

Climates of the West

City	July % of Sunshine	Days of Growing Season	Inches of Rain	Last Frost	First Frost	JAN.	FEB.	MARCH	APRIL	MAY	JUNE	JULY	AUG.	SEPT.	OCT.	NOV.	DEC.	
		During Growing Season				Average Maximum/Minimum Temperature												
Bend, OR	83	91	12"	6-8	9-7	42/17	53/23	50/23	65/28	58/32	78/43	80/43	84/48	69/36	63/31	48/24	43/26	160 days or less
Gt. Falls, MT	80	139	15"	5-9	9-25	29/12	36/17	40/21	54/32	65/41	72/49	84/55	82/53	70/48	59/37	43/26	35/18	
Cheyenne, WY	68	141	15"	5-14	10-2	38/15	41/17	43/20	55/30	65/40	74/48	84/54	82/53	73/43	62/34	47/23	40/18	
Reno, NV	92	155	7"	5-8	10-10	45/18	51/23	56/25	64/30	72/37	80/42	91/47	89/45	82/39	70/30	56/24	46/20	
Ogden, UT	81	155	16"	5-6	10-8	36/18	47/25	48/28	68/42	65/43	88/60	91/62	86/59	78/51	68/41	51/31	43/28	
Pocatello, ID	76	161	23"	4-28	10-6	32/18	46/29	45/30	64/36	62/38	81/48	82/49	86/51	66/43	59/35	42/28	37/26	From 160 to 200 days
Denver, CO	71	171	15"	4-26	10-14	44/14	52/24	54/26	63/39	75/46	76/57	88/61	83/57	81/49	69/37	51/20	49/21	
Centralia, WA	65	173	46"	4-27	10-17	44/31	54/37	52/35	64/39	63/42	74/50	76/50	82/56	68/49	61/42	53/39	46/36	
Pueblo, CO	78	174	12"	4-23	10-14	43/12	57/19	60/24	70/39	81/49	91/57	94/61	88/60	86/49	72/35	59/23	52/19	
Boise, ID	88	177	11"	4-23	10-17	36/21	44/27	52/30	61/36	71/44	78/51	90/58	88/57	78/48	65/39	49/31	39/25	
Santa Fe, NM	78	178	13"	4-24	10-19	36/15	46/22	49/25	61/36	71/43	85/54	85/57	84/58	78/49	69/70	55/30	47/17	
Yakima, WA	82	190	8"	4-15	10-22	36/19	46/25	55/29	64/35	73/43	79/49	88/53	86/51	78/44	65/35	48/28	39/23	
S.L. City, UT	84	192	15"	4-13	10-22	37/18	43/23	51/28	62/37	72/44	81/51	91/60	90/59	80/49	66/38	50/28	39/21	
Albuquerque, NM	76	198	8"	4-13	10-28	47/23	53/27	59/32	70/41	80/51	89/60	92/65	90/63	83/57	72/45	57/32	47/25	
Eugene, OR	60	205	43"	4-13	11-4	46/33	52/35	55/36	61/39	68/44	74/49	83/51	81/51	76/47	64/42	53/38	47/36	From 200 to 250 days
Santa Rosa, CA	65	207	30"	4-10	11-3	59/34	68/39	66/37	76/42	73/44	84/51	88/50	87/53	82/52	79/46	70/40	61/43	
Las Vegas, NV	87	239	4"	3-16	11-10	56/33	61/37	68/42	77/50	87/59	97/67	101/75	101/73	95/65	81/53	66/41	57/34	
Tucson, AZ	78	245	11"	3-19	11-19	63/38	67/40	71/44	81/50	90/57	98/66	98/74	95/72	93/67	84/56	72/45	65/39	
Fresno, CA	96	250	10"	3-14	11-19	55/36	61/39	67/41	74/46	83/52	90/57	98/63	96/61	91/56	80/49	66/41	55/37	
Eureka, CA	52	253	40"	3-10	11-18	53/41	54/42	54/42	55/44	57/48	60/51	60/52	61/53	62/51	60/48	58/45	58/43	
Seattle, WA	63	255	36"	3-14	11-24	45/35	50/37	53/38	59/42	66/47	70/52	76/56	74/55	69/52	62/46	51/40	47/37	From 250 to 300 days
Portland, OR	69	263	34"	3-6	11-24	54/39	59/37	55/37	68/39	65/45	79/52	84/53	89/59	73/52	65/46	54/37	52/40	
Riverside, CA	81	265	10"	3-6	11-26	65/42	76/44	67/42	78/48	72/51	88/57	96/60	92/63	87/57	84/53	79/45	69/47	
Marysville, CA	90	273	21"	2-21	11-21	51/35	66/40	66/41	80/50	74/50	94/61	95/60	94/61	86/57	80/51	67/42	58/43	
Red Bluff, CA	96	274	22"	3-6	12-5	54/37	59/40	64/42	72/47	81/54	89/62	98/67	96/64	91/60	78/52	64/43	55/38	
Bakersfield, CA	95	277	6"	2-21	11-25	52/37	63/41	69/44	75/50	84/56	91/62	99/69	96/67	91/62	80/53	68/44	57/38	
San Jose, CA	72	299	14"	2-10	1-6	57/38	65/44	63/42	73/47	68/48	79/56	82/56	80/58	79/55	74/51	66/45	60/46	
Phoenix, AZ	94	304	7"	2-5	12-6	65/38	69/41	74/45	84/52	93/60	101/68	105/77	102/76	98/69	88/57	75/45	66/38	300 days or more
Sacramento, CA	97	307	17"	2-6	12-10	53/37	59/40	64/42	71/45	79/50	86/55	93/57	91/57	88/53	77/49	64/42	58/38	
Pasadena, CA	87	313	19"	2-3	12-13	67/45	76/48	68/43	75/50	71/50	82/57	84/60	88/63	84/59	82/56	79/51	69/50	
Santa Barbara, CA	65	331	17"	1-22	12-19	66/42	71/44	66/44	70/49	69/52	71/55	75/57	77/61	76/58	74/55	76/47	68/51	
Palm Springs, CA	65	334	31"	1-18	12-18	70/42	83/47	77/44	91/55	87/56	106/69	110/75	106/75	100/67	93/61	82/48	72/47	
San Francisco, CA	66	356	21"	1-7	12-29	56/46	59/48	60/48	61/49	62/51	64/53	64/53	65/54	69/55	68/55	63/51	57/47	
Los Angeles, CA	82	359	14"	1-3	12-28	66/47	68/48	69/50	70/53	73/56	76/59	83/63	84/64	82/63	78/59	73/52	68/48	
San Diego, CA	67	365	9"	—	—	65/46	66/48	68/50	68/54	69/57	71/60	75/64	77/65	76/63	74/58	70/51	66/47	

Mexico and Arizona, the frequency of summer temperatures in the 100° range is fairly high, and desert winds must be considered. Here, the end of the growing season is not abrupt, and a fall garden is usually the better bet.

Growing Season: 200 to 250 Days

The valleys north and south of San Francisco are a part of this category. Partially influenced by the ocean, summer temperatures here are consistently higher than neighboring coastal valleys. Warm-weather crops (corn, tomatoes, peppers, eggplant, etc.) thrive with this season. Normal winters are mild enough for a fall and winter garden of hardy crops.

Growing Season: 250 to 300 Days

The Portland area has a growing season of 260 days with the last frost of spring in early April and first frost of fall in late November. July temperatures average only 67°

Around Tacoma and Seattle, the growing season is about 250 days. The last frost of spring is usually March 13 or 14, and the first frost of fall can be as late as November 18 to 24. But summer temperatures are low, with a July average of only 63°. From June through August, this area receives less than 60 percent of possible sunshine. It's a great climate for the cool-season vegetables, but only the early varieties of corn and tomatoes are sure to ripen. The length of the season partly offsets the coolness of the season.

In California, the cities of Fresno and Merced are within this season range. The long seasons, together with the high summer temperatures and 95 percent of the possible sunshine, make the ideal climate for the high-sugar melons such as crenshaw, honeydew, casaba, and Persian. Cool-weather crops in these areas are grown in the spring. Low temperatures and ground fog frequently discourage winter gardening.

Growing Season: 300 Days or More

The coastal areas of northern and southern California and the inland valleys of southern California are blessed with this long growing season.

Coastal areas are more or less under the direct influence of the ocean fogs. South of San Francisco is lettuce and artichoke country, with most of the cool-weather crops thrown in for good measure. Possible sunshine for

the summer in San Francisco is 69 percent.

The flow of marine air is not uniform. The fog cover is broken by land forms, creating several consistently open areas where summer temperatures increase and warm-weather vegetables are grown.

The inland valleys of southern California make great vegetable-growing country. The area could be divided into as many as five separate zones, varying from one with marine influence to one with more nearly desert climate. Most of the areas enjoy more than 300 frost-free days.

Southern California coastal areas, from Santa Barbara to San Diego, have nearly all-year vegetable gardening. (The lima bean was once the indicator of this climate.) The pattern of marine air flow is not uniform. Summer warmth will vary due to local fog patterns.

VEGETABLE CLIMATES OF THE MIDWEST, THE NORTH, AND THE NORTHEAST

There may be a world of difference between Duluth MN and Berlin NH, but the problems of the short growing season are much the same.

The short-season gardeners have two things in their favor compared to gardeners in long-season areas. Their summer growing days are longer in length than those of their southern neighbors, and the growing season starts in high gear. All of the advance seeding indoors in peat pots and frost protection are a part of gardening in short-season areas.

Growing Season: 150 Days or Less

Many vegetables fit this climate. Look for short-season varieties of long-season crops such as sweet corn, tomatoes, and melons. Consider hot-caps, clear plastic row covers, and black plastic mulch as extenders of the growing season. Lettuce, cabbage, and other cool-season crops do well through the summer here.

Gardeners where seasons are 120 days find themselves pressed in a fairly short time period between spring and fall frost. Cool-season crops do well through the summer. Potatoes and rutabagas are among the vegetables that like it cool and thrive in this climate. Check the seed racks and catalogs for early, short-season varieties of the long-season crops, such as midget muskmelons that ripen in 60 days, and early 'Sunglow' corn that matures in 62 days.

To make the most of a growing season of about 150 days, give the warm-season crops a headstart by growing from seed indoors or by buying transplants to set out as soon as warm weather arrives. The early varieties of all long-season crops should be selected. Nights are cool enough to permit full-season crops.

Growing Season: 150 to 170 Days

In this zone, the additional growing days extend the season into October. However, the normal (if there is a normal) last frost date is in the first week in May. The best way to extend the harvest period is to start seeds in greenhouses, or indoors in peat pots, or let your nursery be your greenhouse. Take full advantage of the 160-day growing season by planting lettuce, endive, turnips, broccoli, Chinese cabbage, and kale in mid-summer for fall and early winter harvest.

Growing Season: 170 to 190 Days

Here, vegetable gardeners look to October harvests and beyond. Carrots planted in mid-July will be large enough for storage in the ground (with mulch covering) as needed.

April 20 usually signals the last frost of spring where growing seasons are 180 days. Inland areas have a fairly high percentage of clear days whereas coastal portions see less sun. The amount of rainfall varies greatly throughout this climate zone. Parts of the western area receive only 20 inches of rain annually, while areas along the east coast receive twice that amount.

Growing Season: 190 Days or More

Here, vegetable gardeners extend the harvest season by a late planting of cool-weather crops for late fall and winter harvest. For example in central Ohio, these vegetables are planted as late as August 1: bush snap beans, beets, endive, kale, leaf lettuce, head lettuce, and radishes.

These are planted as late as August 15: mustard, collards, and turnips. The winter crop of spinach is seeded September 1.

The warmest areas in this climate zone are those along the southern borders of Missouri, Kentucky, and Virginia, where the growing season is about 200 days. This is, of course, long enough to permit summer plantings of snap beans, carrots, sweet corn, and tomatoes, as well as fall plantings of the cool-season crops.

Climates of the Midwest and North

City	July % of Sunshine	Days of Growing Season	Inches of Rain	Days of Rain	Last Frost	1st Frost	JAN.	FEB.	MARCH	APRIL	MAY	JUNE	JULY	AUG.	SEPT.	OCT.	NOV.	DEC.	
Duluth, MN	67	125	15"	47	5-22	9-24	18/−1	21/0	31/11	47/27	61/38	70/47	77/54	75/53	65/44	55/35	35/19	22/6	150 days or less
Bismarck, ND	75	136	10"	42	5-11	9-24	20/−2	23/2	37/17	55/32	67/42	76/52	85/59	83/55	72/45	59/33	38/18	26/5	
Huron, SD	77	149	12"	44	5-4	9-30	25/2	28/7	43/20	60/33	72/44	81/55	90/61	87/59	77/48	64/36	44/21	30/9	
Rapid City, SD	73	150	11"	44	5-7	10-4	31/9	36/12	43/20	57/32	67/43	76/52	86/59	85/57	74/47	62/36	47/24	37/14	
Sioux Falls, SD	75	152	16"	45	5-5	10-3	24/4	30/9	42/22	59/34	71/45	80/56	88/62	85/60	75/49	63/37	43/21	29/9	150 to 170 days
Marquette, MI	67	159	16"	60	5-13	10-19	26/13	27/13	33/20	47/32	59/41	70/51	76/58	74/57	66/50	55/41	39/28	29/19	
North Platte, NE	76	160	14"	45	4-30	10-7	37/11	42/16	50/23	62/35	72/46	82/56	89/62	88/60	78/49	67/36	51/23	40/15	
La Cross, WI	72	161	19"	54	5-1	10-8	25/6	29/10	41/22	57/36	70/48	79/58	85/63	82/61	73/52	61/40	43/26	29/12	
Green Bay, WI	64	161	16"	55	5-6	10-13	25/8	26/9	37/26	52/32	65/44	75/54	81/59	79/57	70/50	58/39	41/26	27/13	
Minneapolis/ St. Paul, MN	70	166	17"	55	4-30	10-13	22/2	26/5	37/18	56/33	69/45	78/55	84/61	81/59	72/48	61/37	41/22	27/9	
Sioux City, IA	71	169	18"	52	4-27	10-13	28/9	32/13	43/24	60/38	73/50	83/60	90/65	87/63	78/53	67/41	47/25	35/15	170 to 190 days
Des Moines, IA	75	175	20"	55	4-24	10-16	29/11	32/15	43/25	59/38	71/50	81/61	87/65	85/63	77/54	66/43	47/28	34/17	
Fort Wayne, IN	71	179	20"	59	4-24	10-20	33/19	35/21	45/28	58/38	70/48	79/58	84/63	82/61	75/54	63/43	47/32	35/22	
Peoria, IL	69	181	21"	55	4-22	10-20	34/18	37/20	47/28	61/40	72/51	82/61	87/65	85/63	78/55	67/44	49/31	37/21	
Detroit, MI	66	182	18"	58	4-21	10-20	34/19	35/19	44/26	58/37	70/47	80/57	85/61	84/60	76/52	65/43	49/32	37/23	
Springfield, IL	73	186	21"	59	4-20	10-23	36/20	41/22	50/30	65/42	76/52	85/62	90/66	87/63	80/55	69/49	52/32	40/24	
Pittsburgh, PA	63	187	20"	65	4-20	10-23	37/21	38/21	46/27	60/38	71/48	80/57	83/61	82/60	76/53	64/42	50/32	38/23	
Milwaukee, WI	70	188	18"	61	4-20	10-25	29/14	32/17	41/26	53/36	64/45	75/55	81/64	79/60	72/53	60/42	45/30	33/19	
Parkersburg, WV	63	159	21"	66	4-16	10-21	43/26	44/27	54/34	65/43	75/52	83/62	86/65	84/64	79/57	68/46	54/36	44/28	
Omaha, NE	81	189	23"	58	4-14	10-20	32/13	37/17	48/28	63/41	73/52	83/62	89/68	86/65	78/56	67/44	49/29	36/19	
Grand Rapids, MI	67	190	16"	52	4-23	10-30	31/17	32/16	42/24	57/35	69/45	79/56	85/60	83/58	74/50	63/39	46/28	34/20	190 days or more
Chicago, IL	70	192	21"	61	4-19	10-28	33/19	35/21	44/29	57/40	69/51	80/61	84/67	82/66	75/57	63/47	47/33	36/22	
Cincinnati, OH	76	192	21"	65	4-15	10-25	41/25	43/27	53/34	64/43	74/53	83/62	87/66	85/64	80/58	68/46	53/36	42/27	
Charleston, WV	63	193	22"	71	4-18	10-28	46/26	49/28	57/33	68/42	77/50	85/59	87/63	86/62	81/56	71/44	57/35	48/29	
Indianapolis, IN	74	193	22"	60	4-17	10-27	37/20	40/23	50/34	62/40	73/50	83/60	88/64	86/62	79/55	67/44	51/32	39/23	
Cleveland, OH	69	195	20"	68	4-21	11-2	36/21	34/21	45/28	57/37	70/48	80/53	85/63	83/61	76/55	64/44	49/34	38/24	
Columbus, OH	71	196	21"	64	4-17	10-30	39/22	40/23	50/30	61/39	73/49	82/52	86/63	84/60	78/55	66/43	51/33	39/24	
Lexington, KY	64	198	24"	66	4-13	10-28	43/25	45/27	55/34	66/43	75/53	84/62	87/66	86/64	81/58	69/47	54/36	45/27	
Topeka, KS	66	200	26"	57	4-9	10-26	39/19	44/23	54/31	66/43	75/52	85/63	91/68	89/66	81/57	71/45	54/31	43/22	
Springfield, MO	70	201	26"	59	4-12	10-30	43/24	47/27	55/33	66/45	75/54	85/63	90/67	90/66	83/58	72/47	56/34	46/27	
St. Louis, MO	71	206	23"	60	4-9	11-1	40/23	44/25	53/32	66/44	75/53	85/63	89/67	87/66	81/58	70/47	54/34	43/26	
Kansas City, MO	76	207	28"	61	4-6	10-30	40/23	45/27	43/34	66/46	75/56	85/66	92/71	90/69	83/60	72/47	55/35	44/28	
Wichita, KS	84	210	24"	56	4-5	11-1	41/23	48/26	56/34	67/45	75/54	86/65	92/69	92/68	83/60	71/49	55/35	45/26	
Evansville, In	77	216	24"	73	4-2	11-4	43/26	47/28	57/36	68/46	77/54	85/64	89/67	87/65	82/59	71/48	56/36	46/28	
Louisville, KY	73	220	25"	68	4-1	11-7	44/26	46/28	56/35	67/45	77/54	85/63	89/67	88/65	82/58	71/47	55/36	46/28	

Climates of the Northeast

City	July % of Sunshine	Days of Growing Season	Inches of Rain	Days of Rain	Last Frost	1st Frost	JAN.	FEB.	MARCH	APRIL	MAY	JUNE	JULY	AUG.	SEPT.	OCT.	NOV.	DEC.	
	During Growing Season						Average Maximum/Minimum Temperature												
Berlin, NH	46	109	12"	41	5-29	9-15	28/6	35/7	37/17	52/31	65/40	74/49	79/53	77/51	69/44	58/34	45/27	31/11	150 days or less
Greenville, ME	60	116	14"	43	5-27	9-20	24/4	29/7	35/14	49/28	63/38	72/48	78/52	75/50	66/43	54/34	41/25	27/10	
Caribou, ME	56	125	15"	54	5-19	9-21	20/1	22/3	32/14	45/28	60/39	69/49	75/54	73/52	64/43	52/34	37/23	24/7	
St. Johnsbury, VT	52	127	14"	46	5-22	9-23	29/7	34/11	40/19	56/32	69/42	78/51	82/55	80/53	72/47	60/36	46/28	32/13	
Pittsfield, MA	60	138	19"	49	5-12	9-27	31/13	32/13	40/22	53/33	66/43	75/51	79/56	78/55	70/47	60/37	46/29	34/17	
Canton, NY	59	140	15"	48	5-9	9-26	27/8	31/12	38/20	56/35	68/44	77/54	81/58	79/56	71/49	60/39	46/30	32/15	
Concord, NH	56	142	14"	48	5-11	9-30	32/9	33/10	43/21	56/30	69/41	78/50	83/55	80/53	72/45	62/34	48/26	34/14	
Burlington, VT	62	148	16"	58	5-8	10-3	28/8	28/8	39/20	53/32	67/43	78/53	82/58	80/56	71/48	59/38	44/28	31/14	
Worcester, MA	60	148	18"	50	5-7	10-2	31/17	33/17	41/25	54/36	66/44	74/55	79/61	77/59	70/52	60/41	47/31	34/20	
Altoona, PA	60	151	22"	62	5-6	10-4	35/20	39/22	45/26	62/39	71/46	78/55	83/59	81/58	75/51	64/41	50/31	38/22	150 to 170 days
Watertown, NY	67	151	16"	50	5-7	10-4	28/10	31/14	38/23	55/36	65/46	75/56	80/61	79/59	71/51	60/41	47/32	33/17	
Binghamton, NY	66	154	18"	56	5-4	10-6	30/17	31/17	39/24	53/34	65/45	73/54	78/58	76/57	69/50	59/41	45/31	33/21	
Bangor, ME	62	156	17"	47	5-1	10-4	28/12	32/14	38/22	52/34	64/44	72/52	78/58	77/56	68/49	57/39	46/31	32/17	
Williamsport, PA	60	164	19"	61	5-3	10-13	36/21	38/21	47/28	60/39	72/49	81/58	85/62	83/60	75/53	64/42	50/33	38/23	
Syracuse, NY	67	168	17"	60	4-30	10-15	32/16	32/17	40/25	55/37	68/47	78/57	82/62	81/60	72/52	61/42	47/33	35/21	
Portland, ME	65	169	16"	54	4-29	10-15	32/12	34/12	41/22	53/32	64/42	73/51	80/57	78/55	70/47	60/37	48/29	35/16	
Albany, NY	64	169	17"	58	4-27	10-13	31/14	33/15	42/24	57/36	70/46	79/56	84/60	81/58	73/50	62/40	48/31	35/18	
Bridgeport, CT	65	173	18"	49	4-26	10-16	37/22	37/21	45/29	55/37	67/48	76/58	82/64	80/63	74/56	64/45	52/35	40/25	170 to 190 days
Scranton, PA	63	174	19"	63	4-24	10-14	34/22	35/21	44/29	57/39	69/50	77/59	82/63	79/61	71/53	60/42	48/33	35/24	
Buffalo, NY	69	179	17"	60	4-30	10-25	31/18	31/17	39/24	53/34	66/44	75/54	80/59	79/58	72/51	60/41	47/32	34/21	
Hartford, CT	62	180	21"	59	4-22	10-19	36/18	38/18	47/27	60/36	72/47	81/57	86/62	83/60	76/52	65/41	51/31	39/20	
Wilmington, DE	65	191	23"	56	4-18	10-26	42/25	43/25	53/32	63/40	75/51	83/60	87/65	85/63	79/57	67/45	55/36	44/26	190 days or more
Winchester, VA	62	193	22"	63	4-17	10-27	43/26	47/28	52/32	68/44	76/52	83/60	88/65	86/63	80/56	69/46	56/36	45/28	
New Haven, CT	66	195	21"	54	4-15	10-27	37/21	37/21	46/28	55/37	66/47	75/56	80/62	79/61	73/54	63/43	52/34	40/24	
Providence, RI	63	197	21"	62	4-13	10-27	37/21	37/20	45/29	55/37	66/47	75/56	80/62	79/60	72/53	62/43	51/34	39/24	
Harrisburg, PA	68	204	21"	68	4-9	10-30	39/24	41/24	50/30	63/40	74/51	73/60	87/65	85/63	78/56	67/45	52/34	41/25	
Charlottesville, VA	62	209	28"	66	4-11	11-6	46/28	49/30	54/35	69/47	77/55	83/63	88/68	86/66	81/60	70/50	58/38	47/30	
Boston, MA	65	217	23"	73	4-5	11-8	37/23	37/23	45/31	56/40	68/50	76/60	82/65	80/63	73/57	63/47	52/38	40/26	
Trenton, NJ	67	218	25"	67	4-4	11-8	40/26	41/26	49/32	61/42	72/52	81/61	85/67	83/65	76/58	66/48	54/38	42/28	
Newark, NJ	65	219	24"	69	4-3	11-8	40/25	41/25	49/32	61/42	72/52	81/61	86/66	84/65	77/58	66/47	54/37	42/27	
New York City, NY	65	219	24"	69	4-7	11-12	40/27	40/26	48/33	60/43	71/53	80/62	85/68	83/67	77/60	66/50	54/40	42/30	
Washington, DC	63	225	26"	66	3-29	11-9	44/29	46/29	54/36	66/46	76/56	83/65	87/69	85/68	79/61	68/50	57/39	46/30	
Atlantic City, NJ	65	225	27"	65	3-31	11-11	43/27	43/26	50/32	60/42	71/51	79/61	84/66	82/65	76/58	67/48	56/38	45/28	
Philadelphia, PA	62	232	26"	70	3-30	11-17	40/24	42/25	50/32	63/41	73/52	82/60	86/65	84/63	77/56	67/44	54/35	42/25	
Baltimore, MD	64	238	26"	68	3-28	11-19	44/25	46/26	54/32	66/43	76/53	83/61	87/66	85/65	79/58	68/46	56/34	46/26	

CELERY
Utah Ju

PLANNING YOUR VEGETABLE GARDEN

A major key to success is planning your garden. Information on succession plantings and crop rotation will help you estimate your harvests and make the best use of your garden space.

PLANNING FOR A CONTINUOUS HARVEST

Vegetable gardens bring many joys, but they also bring problems. One of the most typical (and most frustrating) occurs when everything ripens at once—a classic illustration of the perils of "too much too soon." It isn't easy to plan for a continuous harvest of fresh vegetables, especially for the beginner. And the limited-space gardener has a thornier challenge than the farm-space gardener.

If you're fortunate enough to have plenty of room, you can block out space for the spring garden and leave some space empty in readiness for the summer garden. This way, you can plan and plant the summer garden without interference from the spring garden (which you've already harvested).

But a city gardener with a 20 × 30 area needs to plan for both the spring and summer gardens (and also for the fall-winter gardens, in long-season areas). Clutching a dozen packets of seeds, each enough to plant 50 or 100 feet of rows, the small-space gardener waxes optimistic and pictures row upon row of beautiful, healthy plants.

But although it might seem as if 20 or 30 heads of lettuce couldn't possibly produce enough salads for the whole family, when the 30 heads all mature within a 10-day period, you know you planted *more* than enough.

A dozen cabbage plants don't seem

Lettuce varieties vary in form and color. In these raised beds, the gardener has planned a "show-off" planting of lettuce varieties with leeks in the foreground.

excessive in their little nursery trays. But they'll make 30 to 40 pounds of cabbage when mature. And that's a lot of sauerkraut.

Garden planning—making the best use of the garden space you have—is the best way to avoid such situations.

Length of Harvest Period

Fortunately, not all vegetables have short harvest periods. If you choose two or three of the short-harvest vari-

eties, you can extend the harvest period. Check the "days to maturity" of the varieties—for instance, there are early, mid, and late cabbages.

These vegetables have long harvest periods:

—Green beans begin producing after 50 to 60 days and continue until frost.

—Eggplant begins in 60 to 70 days and lasts until frost.

—Peppers begin after 60 to 80 days and last until frost.

—Summer squash is harvestable after 50 to 60 days and keeps going until frost.

—Tomatoes begin anywhere from 50 to 90 days after planting and last until frost.

In contrast, a vegetable like corn is picked within a two-week time. And the radish harvest lasts only about two weeks. The shorter the duration of the harvest, the more important small, successive plantings are.

Some root crops will last a long time in the soil, including carrots, beets, parsnips, and salsify. Of these, carrots and beets provide a succession of harvests: You can begin harvesting when they are baby-sized and continue up to maturity. Then they will last in the soil for periods of weeks or months, depending upon the time of year and the storage conditions.

Since leaf vegetables such as leaf lettuce and Swiss chard can be picked a leaf at a time, as needed, they have long harvest periods. Depending upon your climate and the time of year, this period may last several weeks.

Perennial vegetables—asparagus, rhubarb, and Jerusalem artichoke (same as sunchoke)—come back year after year. Measure their duration of harvest in seasons.

Garden planning can be tricky, but when necessary you can gain time, fill in unanticipated gaps after harvest, or even plan totally with transplants from your local nursery or garden center. Transplants are also preferred for some crops, as shown in the chart on pages 48 and 49.

The National Garden Bureau (an educational service of the seed industry) suggests five major considerations to keep in mind when deciding what to plant where:

—Do you and your family like the vegetable?

—How many days are required from planting to harvest?

—Does the vegetable prefer cool or warm weather?

—How large do the plants grow?

—How many plants of each kind will you need to feed your family?

Succession Plantings

The way to a continuous harvest is to make successive plantings of small quantities. For example, if you first plant lettuce in March, you can make another planting in April, and another in May. Successive plantings of snap beans, four to six weeks apart, will give you fresh beans for five months or more. But to get a continuous harvest of a vegetable, you must plant the second planting before you harvest the first.

Early Spring
Plant as soon as ground can be worked in spring: Broccoli plants, Cabbage plants, Endive, Kohlrabi, Lettuce, Onion sets, Parsley, Peas, Radishes, Spinach, Turnips.

Mid-Spring
Plant these at time of the average last killing frost: Carrots, Cauliflower plants, Beets, Onion seeds, Parsnips, Swiss Chard. Plant two weeks later: Beans, Corn, Potatoes, Early Tomato seeds.

GARDEN PLANTING SCHEMES

We've imagined a 25- × 30-foot garden area and made plans for it that will bring a succession of crops throughout the growing season. A small-space gardener can either use this scheme as is or adapt it.

Divide the area into three sections, each approximately 10 feet wide. Plant the first block with early, cool-weather crops such as carrots, beets, spinach, onions, lettuce, radishes, turnips, early cabbage, and early potatoes. If you allow for proper spacing between rows (see chart, page 48), this planting will use a 10-foot depth of that first block. Plant these vegetables in spring, as soon as the soil is workable.

In the second 10- × 10-foot block in the next section, plant the same vegetables, but a month later. Or you could try other vegetables—for example,

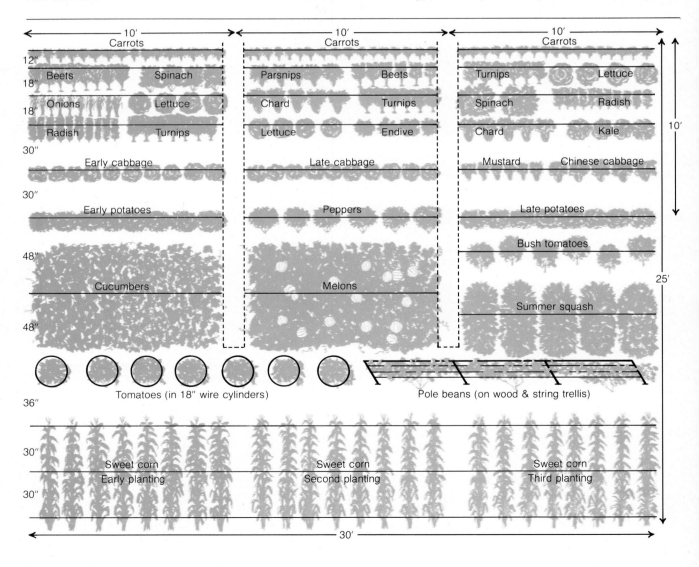

Early Summer
Plant when soil and weather are warm: Lima Beans, Cantaloupe, Celery plants, Crenshaw melons, Cucumbers, Eggplant plants, Pumpkins, Pepper plants, Potatoes for winter, Squash, Tomato plants, Watermelons.

Mid-Summer-Fall
Plant in late June or early July: Beets, Broccoli, Cabbage, Cauliflower, Kohlrabi, Lettuce, Radishes, Spinach, Turnips.

easier than you might think—planting times are determined by only four factors: when the soil is workable in spring; the date of the last spring frost; the date of the first fall frost; and the average time the soil becomes frozen, if ever. Most of these dates can be deduced from the charts on pages 28-33. Next, look up the vegetables you would like to grow, and note when they should be planted and the average number of days to harvest. By combining these figures, you'll know what to plant when, about when it will be ready to harvest, and when to replant for a steady harvest. Once you've charted your plantings in this way, you can make adjustments easily that will lengthen the harvest season.

parsnips, chard, or endive in place of the spinach, onions, and radishes.

Plant the third block with those crops that will mature in fall. Turnips, lettuce, spinach, radishes, chard, kale, mustard, and Chinese cabbage are some cool-weather crops that will benefit by being planted after the hottest days of summer and will mature in fall.

Plant the rest of the garden space with warm-season crops, after the soil is thoroughly warm. These crops include cucumbers, melons, summer squash, tomatoes, corn, pole beans, and peppers. Plant the tomatoes in 18-inch wire cylinders. Train pole beans along a wood-and-string trellis. Plant the corn in three blocks of three rows, making three plantings two weeks apart.

Succession Charting
When you chart a succession of plantings on paper, the goal of a long harvest season comes within reach. It's

Raised Beds
You can also use raised beds to lengthen the vegetable-growing season in most parts of the country. In the spring, they warm earlier than the surrounding soil, since air circulation is freer in looser soil. (See page 12 for more on raised beds.)

A Typical Garden Calendar
You can plan the succession planting and harvest in your garden by making a calendar like this of your area. This one is for the Delaware area, taken from the University of Delaware Extension Service Bulletin 55.

| Plant | Harvest |

Mar	Apr	May	Jun	July	Aug	Sep	Oct	Nov
		Snap beans			Snap beans			
		Lima beans			Lima beans			
Cabbage			Cabbage					
				Cabbage		Cabbage		
		Cucumbers		Cucumbers				
Carrots			Carrots					
				Carrots		Carrots		
Beets			Beets					
				Beets		Beets		
				Broccoli		Broccoli		
				Cauliflower		Cauliflower		
		Cantaloupe			Cantaloupe			
Lettuce		Lettuce			Lettuce	Lettuce		
Onions		Green Onions		Onions				
Peas		Peas						
		Peppers		Peppers				
Radish		Radish				Radish	Radish	
Spinach		Spinach			Spinach		Spinach	
		Sweet Corn		Sweet Corn				
		Squash	Squash					
		Winter Squash				Winter Squash		
		Tomatoes		Tomatoes				
Turnips			Turnips		Turnips		Turnips	
		Watermelon		Watermelon				

Tomatoes can be well trained for containers. Here, staked in a tub, the 'Atom' tomato performs handsomely.

THE GARDENER WITHOUT A GARDEN

All of us might like to have a large, country-style garden, but most people don't. Nevertheless, small-space gardeners are a determined breed. "Gardeners will garden," a veteran smallspacer once said. "It's that simple." Productive vegetable gardens have been grown on rooftops, balconies, decks, and other nontraditional locations.

Of necessity, the gardener without a vegetable garden can neither overplant nor waste the harvest. No time is spent guessing how much a 20-foot row of beets will produce; instead, seeds are simply sown, a box or two at a time.

Some determined gardeners combine resources with their friends and neighbors to make a community garden, an undertaking that's simultaneously practical, social, and educational—practical because it makes good use of the vacant lots found in urban areas, and lets the gardeners grow food close to where it is used; social because it offers a great way to get to know your neighbors; and educational because of the gardening knowledge that's shared.

Vegetables in Containers

Versatiiity is the hallmark of the container gardener. On balconies, decks, or patios, vegetables can grow in boxes, tubs, bushel baskets, cans, and planters of all shapes and sizes. Any of these containers will work—the depth of the box is what's critical.

The major considerations to keep in mind are portability, frequency of watering, and fertilizing. The shallower the box, the more frequent is the need for water and fertilizing. For the following vegetables, the boxes should have these minimum depths:

4 inches deep—Lettuce, turnips, radishes, beets, and all the low-growing herbs.

6 inches deep—Chard, kohlrabi, short carrots such as 'Baby Finger', and the root crops listed above.

8 inches deep—Bush beans, cabbage, eggplant, peppers, and bush cucumbers.

10 inches deep—Cauliflower, broccoli, and brussels sprouts.

12 inches deep—Parsnips, salsify, long-rooted carrots, and tomatoes.

The choice of which vegetables to plant in container gardens also depends on which ones give the highest return per square foot of space—in other words, those that can be spaced most closely in the row. (See "Planting Chart," page 48.) Some vegetables in this group are carrots, beets, chives, leaf lettuce, mustard, green onions, radishes, and turnips.

For a continuous harvest, you might plant a total of 6 boxes—2 early; 2 more three weeks later; and the last 2 two weeks after that.

How vegetables grow in your climate will dictate your choice of late spring and fall plantings. Box plantings make it easy to think of harvests in terms of the number of meals instead of the total quantity of plants.

Compact Varieties

Gardens that cannot grow out can grow up. A fence 5 feet high and 20 feet long provides a whopping 100 square feet of growing space. It can be covered with vines of beans, cucumbers, tomatoes, squash, or gourds without infringing on most of the ground space. Pole beans yield more per plant than bush beans; upright planters produce a row's worth of strawberries, lettuce, or herbs.

The needs of small-space gardeners have not gone unheeded by vegetable plant breeders; they have come up with many "miniature" and "bush" varieties of favorite vegetables. If you haven't the room for vining cucumbers, try 'Pot Luck'. The full-sized cukes it produces are almost as big as the whole plant. Also consider the bush-type squashes such as 'Table King', or the new space-saving pumpkin, 'Spirit'.

Our readers have written about planting potatoes in everything from bushel baskets to stacks of old tires. Here, a good harvest results from planting in a plastic garbage container.

2" plastic tubes riddled with ¼" holes for watering.

Black plastic lining held in place by crossed laths on 6" centers.

2"x4" caps.

2"x10" base with drain holes.

2"x10" sides.

Heavy-duty casters

Fill with soil mix. Cut holes through plastic to insert small plants.

Above: Romaine lettuce makes a beautiful display in this roll-around box (details shown). Below: The versatility of containers and plants for containers is clearly shown here. It's obvious that limited space need not prevent a bountiful harvest.

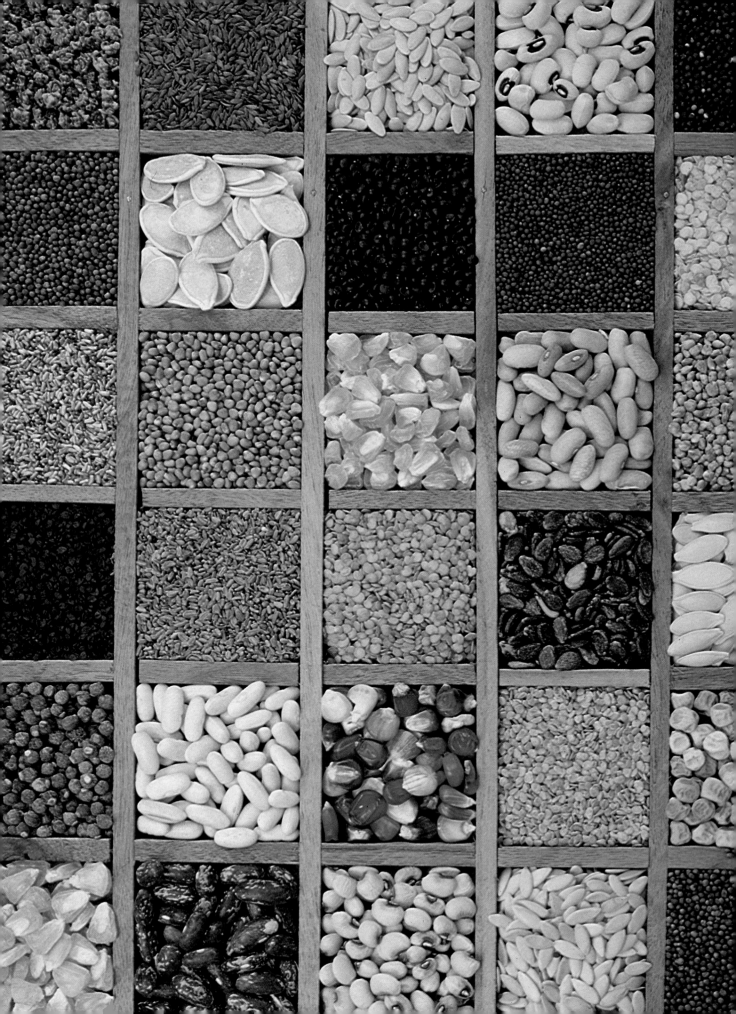

VEGETABLE SEEDING AND TRANSPLANTING

Here is the information you need to get plants off to a good start. Extensive planting charts, tips on starting seeds indoors, data on germination periods and advice on protecting transplants will help you every step of the way.

Seeding

On the face of it, the act of sowing seeds ¼ or ½ or 1 inch deep, covering them with soil, and keeping that soil damp would seem simple enough. But in fact, it doesn't always work.

"For the first time in 13 years," confessed one disgruntled gardener, "absolute failures were experienced. Diseases not before experienced were seen. Plants in excellent health from my greenhouse grew magnificently and then perished. Corn, cabbage, butternut squash, and onions did not even appear from seed."

Even the best of gardeners have sometimes had seeds that failed to germinate.

If you are a first-time gardener, guard against the tendency to kill by kindness. Don't overprepare a seed bed. Don't rake the soil until it is fine as dust. That's a good way to make mud pies—when the soil dries out after watering, it will imprison the seeds in a tight crust.

Don't tenderly cover the seed with loose soil. Tamp the soil down firmly —seeds need contact with it. To firm the soil: use the side of a rake or the flat of a hoe; use a short piece of 2 by 4; or you can simply pat the soil with your hands.

Improving heavy-clay soil: You don't have to redo the soil in the entire garden to improve soil that's heavy clay. Just focus on the row to be planted, adding compost or other organic material to it.

Don't rush the season: Many seeds tend to rot instead of root. This is particularly true of corn and beans (especially limas in cold soil). The warmer the soil, the better the chance such seeds have to succeed.

If you find yourself with spring fever but the soil is still cold, at least choose seeds that tolerate cool soil. (See the planting chart on page 48).

There is no absolute rule that applies to planting all vegetable seeds. You will find recommendations in the planting chart on pages 48-49.

Beauty in the vegetable garden begins here. Seeds from top and left: (row 1) beets, lettuce, cucumber, blackeye peas, kale; (row 2) cabbage, pumpkin, asparagus, turnip, pepper; (row 3) endive, radish, corn, beans, spinach; (row 4) onion, carrot, eggplant, watermelon, squash; (row 5) malabar spinach, 'gold crop' bush beans, ornamental corn, tomato, sugarsnap peas; (row 6) 'silver queen' sweet corn, 'tendergreen' bush beans, purple hull peas, muskmelon, broccoli.

Depth of seed: While there's no exact rule about how deep to plant your seeds, the old rule of thumb is: Plant at a depth equal to four times the diameter of the seed. However, be sure to weigh this rule—and even the specific depths listed in the planting chart—against your own judgment. If the weather is wet or the soil is heavy, plant shallow. If dry weather is expected and your soil is light and sandy, plant deeper.

Several crops—parsnips, carrots, parsley, cress, and (to a lesser degree) lettuce—benefit by very shallow planting, ¼ inch or less.

However, when you plant seeds with a long germination period at this shallow a depth, you need to sprinkle the soil attentively and frequently so that it never becomes dry.

Mulching

Your mulch may be made of organic material (such as leaves, peat moss, straw, manure, sawdust, ground bark, compost, and the like), or manufactured materials (such as polyethylene film, aluminum foil, or paper).

Manufactured materials: Some gardeners cover soil rows with burlap sacks and sprinkle as necessary. With sacks there is always the danger of forgetting them for a few days at emergence time.

Clear plastic makes a good mulch but, again, watch closely for the emergence of seedlings and remove the plastic as soon as the seedlings show.

When asked about mulch cover, Bernard Pollack of Rutgers University answered the question this way:

"I think plastic is the answer. Make a shallow trough and plant seeds at bottom—cover trough with plastic (clear only) at an angle so water can run off. (See Page 44.)

"In New Jersey we call this 'trench culture' and we have had exceptionally good results with direct seeding of hard-to-germinate seeds like tomato, pepper, eggplant. The plastic cover prevents evaporation, heats the soil 20 to 30 degrees higher than air temperatures, and makes the seed germinate fast; and that helps prevent loss from disease. Plastic should be removed after germination to control weeds."

Organic Mulches: An organic mulch poses many advantages, but it needs to be applied at the right time. If applied in early spring it will slow up the natural warming of the soil as spring advances. Thus it will become an insulating blanket, reducing solar radiation into the soil and increasing the chance of frost

hazards. Therefore, summer is the best time to add organic mulch, when the soil is already sufficiently warm.

Mulch thickness: A ⅛-inch mulch of vermiculite, bark, or sawdust will prevent crusting and reduce the need to water frequently. It's a good idea to contain the mulch in a trench to keep it from blowing away.

For established plants apply organic mulches 1 to 2 inches thick for fine materials (for example, sawdust), and up to 4 inches thick for coarse materials (such as straw). If your sawdust is untreated, increase the amount of fertilizer regularly used for the crop by ¼, to reduce the loss of nitrogen from the soil.

Germination of Seeds

Temperature, moisture, and oxygen supply are the three most significant factors that influence germination. In some cases, light constitutes a fourth. Most vegetable seeds will tolerate considerable variation in these factors; however, as extremes are approached (above or below optimum), the germination rate slows, increasing the amount of abnormal seedlings and reducing total germination.

Of all the germination-controlling

A styrofoam picnic box can be converted into a miniature greenhouse. They work very well for seed propagation. Nighttime heat retention is outstanding.

This deep layer of straw keeps the soil cool for good lettuce production. Be sure the straw is seed-free, though, or you may have a problem with weeds.

factors, report J. F. Harrington and P. A. Minges of the University of California, Davis, temperature is the one that has been studied most extensively. (See the chart on page 43, which summarizes much of their work in this field.)

Temperature: This factor affects vegetable-seed germination both directly and indirectly. Above the maximum and below the minimum germination temperatures, no germination occurs. Harrington and Minges say:

"Although lettuce and onion will germinate well at 32 degrees F., the rate of germination is extremely slow —49 and 136 days, respectively. The important point is that these seeds can stand this cold temperature and still be alive when the soil warms, while most other seeds, such as beans and sweet corn, will rot if left long at low temperatures.

"The minimum temperature of germination can be used as an indication of when to plant early crops in the spring. When the soil temperature has risen to the minimum temperature for germination, the seed may be planted with the expectation that the soil will continue to warm to temperatures where germination is more rapid. Planting a given crop before the soil has warmed to the minimum temperature for germination does not produce an earlier crop and may result in a greatly reduced stand.

"Maximum temperatures are for seeds in moist soil with otherwise good germination conditions. During hot weather, soil temperatures near

the surface often exceed these critical temperatures during the day, and in some cases this may be the cause of poor stands. High soil temperatures are most damaging at the time of seedling emergence.

"Even if the seed should germinate in high-temperature soils, the seedlings may die because of heat injury. Carrots and beets are examples of seedlings likely to be damaged this way, because their thin upright seedling growth provides almost no shade

around the stem. A cloddy seedbed which reduces heating by shading the soil surface usually helps prevent excessive seedling loss due to heat."

Moisture. The stored food in a seed occurs in a very concentrated, complex form. Before it can be used, a series of chemical reactions must take place; and for this to happen, moisture must be available. Water serves two functions: It triggers this series of reactions within the seed, and it softens and weakens the seed coat,

MOISTURE AND SEED GERMINATION*

Germinate well in a wide range of soil-moisture conditions:	Germinate best in fairly moist soil:	Germinate only in wet soil:	Germinate best in relatively dry soil:
Cabbage	Beet	Celery	Spinach
Carrot	Endive		Spinach, New Zeland
Corn, sweet	Lettuce		
Cucumber	Lima bean		
Muskmelon	Pea		
Onion	Snap bean		
Pepper			
Radish			
Squash			
Tomato			
Turnip			
Watermelon	*Adapted from Doneen and MacGillivray		

SOIL TEMPERATURES AND VEGETABLE SEED GERMINATION
(degrees Fahrenheit)*

VEGETABLE	MINIMUM	OPTIMUM	MAXIMUM
Asparagus	50	75	95
Bean, lima	60	80	85
Bean, snap	60	85	95
Beet	40	85	95
Broccoli	40	85	95
Cabbage	40	85	95
Carrot	40	80	95
Cauliflower	40	80	95
Celery	40	70	75
Chard, Swiss	40	85	95
Corn, sweet	50	85	105
Cucumber	60	95	105
Endive	32	75	75
Lettuce	32	75	75
Muskmelon	60	95	105
Okra	60	95	105
Onion	32	80	95
Parsley	40	80	95
Parsnip	32	70	85
Pea	40	75	85
Pepper	60	85	95
Pumpkin	60	95	105
Radish	40	85	95
Spinach	32	70	75
Squash	60	95	105
Tomato	50	85	95
Turnip	40	85	105
Watermelon	60	95	105

*Adapted from Harrington and Minges, University of California, Davis

Below: This miniature greenhouse with a dowel handle is a good seed starter. Bottom: The A-frame with a plastic cover is portable and useful as an early spring row cover.

Sowing Carrots

To make the most of your space when sowing carrot seed, sow randomly in a swath 6 to 12 inches wide, and cover with ¼ inch of fine peat moss. Thin seedlings randomly and enjoy the sweet, tender miniature carrots as they grow.

Spacing Seeds

Space small seeds evenly by rubbing a pinch between fingers . . .

or tap them directly from the packet

Sowing Small Seeds

When the small size or color of seeds makes them difficult to see as you're sowing, lay sheets of tissue paper in the trench. The tissue will decompose quickly when covered and watered. Seed tape is also available for many plants.

Trench Greenhouse

To help start hard-to-germinate seeds like tomatoes, peppers, and eggplant, plant seeds in a trench covered with clear plastic. Angle it so water drains off.

thus permitting the growing embryo to break through.

To determine the optimum moisture conditions for germinating vegetable seeds, a series of tests was conducted. The chart, "Moisture and seed germination" on page 43, shows the results.

Oxygen. Since a seed is in a state of suspended animation, its energy requirements are low. As its growth begins, triggered by the presence of moisture, oxygen combines with the stored food to make energy.

Lack of oxygen rarely is the cause of a vegetable seed's failure to germinate. This factor usually is influential only when the pore space in the soil around the seed is saturated with water, such as after a heavy rain. If flooded conditions persist for too long, the seed will be killed.

Some seeds are more sensitive to low oxygen than others, particularly squash, pumpkin, cucumber, gourds, and related plants. On the other hand, celery seed can germinate even if it is completely immersed in water.

Seed Sowing

When most people think about sowing seeds, they think in terms of a single-line row, but you might instead consider the advantages of sowing seeds of carrots, beets, and leaf lettuce in 3- or 4-inch-wide bands. Bands minimize the problem of doing the first thinning. When the same number of seeds normally sown in a foot of row are spaced out 3 to 4 inches wide, there is less chance of tangled, malformed roots. You will have to do some thinning, of course, but much of it will be in pulling baby carrots. The disadvantage of the wide row, however, is that you have to weed by hand.

Commercial carrot-growers equip their planters with a "scatter shoe," which spreads the carrot seeds in a 6-inch band.

You can use this same method for all fine-seeded vegetables that are used in the immature stage—for example, beets and turnips (for their green tops), and leaf lettuce and mustard (for their young leaves).

Spacing Seed in Furrows

Some of the ways to drop seed in furrows are illustrated here. Here are a few more:

Sow small seeds in groups of two to six with a few inches between groups. Some gardeners feel that the seeds help each other up when the soil is likely to crust over, which helps ensure against seed failures.

Mix small seeds and white sand in a salt shaker and shake them down the furrow.

If you prefer your small seeds more precisely spaced, spread single sheets of white facial tissue in the seed furrow. The seeds will show up clearly against the white paper, and you can move them easily with a pencil or toothpick. The tissue will rot away long before the seeds germinate.

Or you can do the same thing using a commercially available aid—a water-soluble plastic that encloses precisely spaced seeds.

Seed Storage

Seed that is stored under ideal conditions will maintain high germination rates—even up to several times longer than most charts on seed storage suggest. However, if you store seed improperly, it may die within a month.

What are proper storage conditions? Basically, the reverse of ideal germination conditions.

Experts Harrington and Minges have this to say about storing seed:

"The two most important factors are the moisture content of the seed, and the temperature. If the moisture content of the seed is low, the seed will keep well under wide differences of temperature. Likewise, if the storage temperature is low, the moisture in the seed, if not extremely high, is not a serious factor causing deterioration.

"If the seed is both dry and cold, then conditions are nearly ideal. To be more specific, seeds of high germination to start with will maintain high germination for ten years or more if dried below 8 percent moisture, then sealed in moisture-proof cans and stored in cold storage of 32 to 40 degrees F. If the humidity of the air within the container is 70 percent or more, the moisture content of most seeds will rise above 20 percent. If the temperature then rises to 80 degrees F. or above, the seeds will be dead within one or two months. It is better to store seeds in an open container unless seed moisture content is below 8 percent."

Weeds

Unfortunately, weed seedlings tend to compete with vegetable seedlings. Every time you work the soil, you inadvertently bring weed seeds up to the surface, where they find excellent conditions for germination. And, ironically, every attempt to cover your crop (whether with clear plastic or a mulch) aids the weeds, as well.

If you're a first-time gardener, you may wonder whether your seed packet didn't, in fact, contain 50 percent weed seeds. Gerald Burke of the Burpee Company says:

"One complaint we run into quite

consistently is the story of the gardener who plants good seed out of the packet, but nothing but weeds come up. I guess this one bugs us as much as anything else. With today's highly developed techniques of cleaning vegetable seed, the chance of weed seed appearing in any of these crops along with the good seed is negligible. There is 1,000 times more weed seed in the soil than is ever going to be in a packet of seed."

TRANSPLANTING

With root crops, beans, peas, corn, pumpkin, and squashes, you sow seeds directly where you want them to grow. But with tomatoes, peppers, and eggplants, you almost always have to set them out as transplants. However, cabbage and its relatives, lettuce, onions, and melons can be started either way.

Not only does growing transplants save time (an important consideration), but it also permits the plant to grow before frost danger is over and before the soil is workable. In this way, growing transplants actually can lengthen the growing season by one to two months.

Plants begun as transplants also avoid some of the hazards common to seedlings—birds, insects, heavy rain, and weeds.

You can buy transplants or grow your own. Which route you choose depends on whether the varieties you want are available, and on whether the idea of starting from scratch appeals to you.

There are different methods of growing from seed to transplant size. Some of them are illustrated below.

To grow successful transplants, you need:

—A disease-free growing medium.
—Warmth and moisture for seed germination.
—Adequate light for stocky growth.
—An adjustment period to ready the indoor plant to outdoor conditions.

There are several ways to meet the above requirements.

The Growing Medium

You can buy many excellent germinating and growing materials at the garden supply centers. Aside from the several kinds of commercial synthetic soils (See Page 10), most supply centers carry some of the ready-to-use planting blocks or cubes. Some gardeners prefer starting transplants in vermiculite.

Starting Seeds

Here are some good ideas for starting seed and handling transplants that we picked up from the backyard gardeners.

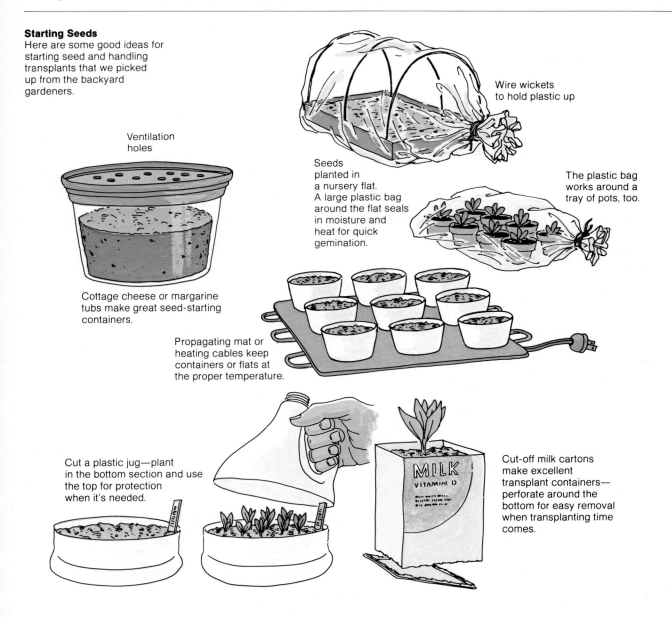

Ventilation holes

Wire wickets to hold plastic up

Seeds planted in a nursery flat. A large plastic bag around the flat seals in moisture and heat for quick gemination.

The plastic bag works around a tray of pots, too.

Cottage cheese or margarine tubs make great seed-starting containers.

Propagating mat or heating cables keep containers or flats at the proper temperature.

Cut a plastic jug—plant in the bottom section and use the top for protection when it's needed.

MILK VITAMIN D

Cut-off milk cartons make excellent transplant containers—perforate around the bottom for easy removal when transplanting time comes.

Seed Starting Containers

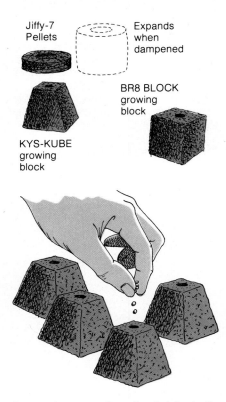

Jiffy-7 Pellets

Expands when dampened

KYS-KUBE growing block

BR8 BLOCK growing block

Sow seeds, two at a time, directly into plastic pots, KYS-KUBES, BR8 BLOCKS, or Jiffy-7 pellets. Water thoroughly and place on a tray in a plastic bag. They'll be ready to transplant when about 6 inches high.

You can, of course, start your seeds in garden soil or compost, but it's a good idea to sterilize them. You can sterilize small quantities by baking the soil in the oven. Coffee cans full of garden soil or compost can be baked for 1½ hours. Or you can place a potato in the soil and consider the soil sterilized when the potato is baked. (Shallow containers may shorten the time involved; in a 350° oven, the soil may reach 180° in only 45 minutes.) As soon as all the soil reaches 180°, the job is done. Don't overcook—toxic substances can be released from the soil itself.

Seeds in Blocks or Cubes
You can find several kinds of these. Some of them are:

Jiffy-7 pellets. This is a compressed peat pellet containing fertilizer that, when placed in water, expands to make a 1¾- by 2-inch container. After they have expanded, place them directly into a container.

BR8 bloks. This is a plant-growing fiber block containing fertilizer. Water the block thoroughly, then place the seed directly in it.

Kys-Kube. A ready-to-use fiber cube containing fertilizer. Water it thoroughly, then place the seed directly into it.

Fiber Pots—trays—strips. These containers are made of peat or other fibrous material. Fill them with synthetic soil mix and place the seeds in the soil. Once the seedlings come up, set the whole container out in the garden or into your planting container. This will keep the roots from being disturbed.

The important thing to remember is that these containers are made to hold moisture well. Once you have

watered them, you can plant and then cover the blocks or cubes with paper (or slip them into a plastic bag) to prevent them from drying out. There should be no need for more watering until the seeds have sprouted.

Seeds in Vermiculite
Moisten the vermiculite and sow seeds about ¼ to ½ inch deep. Pat the vermiculite lightly to firm it around the seeds. Water lightly. Cover with paper, or slip the tray into a plastic bag.

When the seedlings have formed their first true leaves, dig them out carefully and transplant them into containers filled with synthetic soil. With a pencil, make a small hole in the container mix, and set the seedling in the hole; the seed leaves should be ½ inch from the surface. Press the mix firmly around the roots and stem. Water carefully.

One advantage of starting seeds in vermiculite is that you can lift out the seedlings without damaging the roots. Another is that they won't die from "damping off" disease, which is caused by organisms in the garden soil and compost.

Whether you grow your seedlings in synthetic soil, ready-to-use containers, vermiculite, or any combination that suits you, it is very important to give them enough sunlight.

Providing Enough Sunlight
Once the seedlings emerge, keep them in full sunlight—12 hours a day, if possible. Daytime temperatures should range from 70° to 75°F.; nighttime temperatures should be between 60° and 65°F.

It's easy enough to say, "Give seedlings 12 hours a day of full sunlight and temperatures that range from

Transplanting

   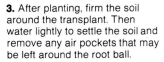

1. Ready the plant. If it's in a peat pot, tear the top edge off so it can't act as a wick and dry out the root ball.

2. If it's in a plastic, fiber, or clay pot, tip it out—don't pull it out by the stem.

3. After planting, firm the soil around the transplant. Then water lightly to settle the soil and remove any air pockets that may be left around the root ball.

4. To make sure the root ball stays moist during the first few critical days, build a small temporary basin, a little larger than the root system.

70° days to 60° nights." And it's even easy enough to follow these directions —if your "indoors" is a greenhouse. But what do you do when your indoors can't meet these conditions? The answer is, you make do. Here are some of the many ways of carrying out the transplant procedure. Some of them are sketched on the facing page.

"Hardening" Transplants

Don't put young plants directly into the open garden from an indoor environment. Instead, take them outside in the daytime and bring them in again at night, if frost is likely. About two weeks before setting them out, gradually expose them to lower temperatures and more sunlight.

There are several ways to protect transplants. You can warm the soil to provide frost protection, but covering the transplants with hotcaps or plastic covers in early plantings offers more advantages. It is equally important to protect the young plants from being ripped about by winds.

When using hotcaps or plastic covers, make sure to provide some ventilation so that the young plants don't get cooked by the heat buildup.

Setting Out

Transplants should go into the soil with a minimum of root disturbance. With peat pots, cubes, and blocks, there will be very little disturbance of the roots; however, to prevent the root ball from drying out rapidly, place all such containers below soil level.

It helps to pack soil down around the rootball. It also helps to "spot" water the rootball in addition to irrigating regularly.

If the soil is wet, plants in plastic or clay pots will tip out more easily.

If a plant is growing in a light soil mix, don't set it in a small hole in heavy clay soil. Instead, dig a larger planting hole and blend organic matter into it. This will prevent the soil texture from changing abruptly.

For ways to protect transplants, see "Good Ideas from Good Gardeners" on page 106.

How to Use the Planting Chart

"Depth to plant seed." A quick look at the fractions and you know that many gardeners plant too deep.

"Number of seed to sow per foot." It's one answer to the question "How thick or thin should I sow seeds?" Our figures give the average of 6 expert seed-sowers—3 pessimists and 3 optimists.

"Distance between plants." First figure is minimum. You get better growth at wider spacing. You cut down on the competition.

"Distance between rows." The minimum distance assumes that space is limited and weeding will be done by hand tools. Wider distance between rows is preferable and if power equipment is used, necessary.

"Number of days to germination."

Above: Varieties of leaf, bunching, and dwarf lettuces are being started here with a shady north exposure, using aluminum foil to reflect and increase the amount of light. Below: The author checks one of his raised beds with a fiberglass coldframe on top. These sweet potatoes were able to thrive where they are not as well adapted. At 10 p.m., temperatures were 2° to 5°F higher than the outside air. Ventilation was necessary in daytime.

Number of days varies by soil temperature. Early spring sowings will take longer than later plantings. We give the range to answer questions like this one: "How long do I wait before I know I have to reseed?"

"Soil temperatures for seed." Seeds that "require cool soil" do best in a temperature range of 50°-65°; that

"tolerate cool soil" in a 50°-85° range; those that "require warm soil" in a 65°-85° range.

"Weeks needed to grow to transplant size." The variation of 4-6, 5-7, 10-12 weeks allows for hot-bed, greenhouse, and window sill, and under grow-lamp conditions. Generally the warmer the growing conditions the

shorter the time to grow transplants. However there must be allowance for a change from indoor to outdoor environment.

"Days to maturity." Figures in this column show the *relative* length of time needed to grow a crop from seed or transplant to table use. The time will vary by variety and season.

Vegetable	Depth to plant seed (inches)	Number of seed to sow per foot	Distance between plants (inches)	Distance between rows (inches)	Number of days to germination	Needs cool soil	Tolerates cool soil	Needs warm soil	Weeks needed to grow to transplant size	Days to maturity	Remarks
Artichoke	½		60	72	7-14		•		4-6	12 mos	Start with divisions preferred.
Asparagus	1½		18	36	7-21		•		1 year	3 years	Sow in spring and transplant the following spring.
Beans: Snap Bush	1½-2	6-8	2-3	18-30	6-14			•		45-65	Make sequence plantings.
Snap Pole	1½-2	4-6	4-6	36-48	6-14			•		60-70	Long bearing season if kept picked.
Lima Bush	1½-2	5-8	3-6	24-30	7-12			•		60-80	Needs warmer soil than snap beans
Lima Pole	1½-2	4-5	6-10	30-36	7-12			•		85-90	
Fava—Broadbean Winsor Bean	2½	5-8	3-4	18-24	7-14		•			80-90	Hardier than the common bean
Garbanzo—Chick Pea	1½-2	5-8	3-4	24-30	6-12			•		105	
Scarlet Runner	1½-2	4-6	4-6	36-48	6-14			•		60-70	Will grow in cooler summers than common beans
Soybean	1½-2	6-8	2-3	24-30	6-14			•		55-85 95-100	Choose varieties to fit your climate. See text.
Beets	½-1	10-15	2	12-18	7-10		•			55-65	Thin out extra plants and use for greens.
Black-eye Cowpea Southern Peas	½-1	5-8	3-4	24-30	7-10			•		65-80	
Yardlong bean Asparagus Bean	½-1	2-4	12-24	24-36	6-13			•		65-80	Variety of Black eye peas. Grow as pole beans.
Broccoli, sprouting	½	10-15	14-18	24-30	3-10		•		5-7*	60-80T	80-100 days from seed.
Brussels Sprouts	½	10-15	12-18	24-30	3-10		•		4-6*	80-90T	100-110 days from seed.
Cabbage	½	8-10	12-20	24-30	4-10		•		5-7*	65-95T	Use thinnings for transplants. 90-150 days from seed.
Cabbage, Chinese	½	8-10	10-12	18-24	4-10		•		4-6	80-90	Best as seeded fall crop.
Cardoon	½	4-6	18	36	8-14		•		8	120-150	Transplanting to harvest about 90 days.
Carrot	¼	15-20	1-2	14-24	10-17		•			60-80	Start using when ½" in diameter to thin stand.
Cauliflower	½	8-10	18	30-36	4-10		•		5-7*	55-65T	70-120 days from seed.
Celeriac	⅛	8-12	8	24-30	9-21	•			10-12*	90-120T	Keep seeds moist.
Celery	⅛	8-12	8	24-30	9-21	•			10-12*	90-120T	Keep seeds moist.
Celtuce—Asparagus Lettuce	½	28-10	12	18	4-10		•		4-6	80	Same culture as lettuce.
Chard, Swiss	1	6-10	4-8	18-24	7-10		•			55-65	Use thinnings for early greens.
Chayote	See text	—	10 ft.	—	See text			•	—	Perennial	Plant whole fruit.
Chicory—Witloof (Belgian Endive)	¼	8-10	4-8	18-24	5-12		•			90-120	Force mature root for Belgian Endive.
Chives	½	8-10	8	10-16	8-12		•			80-90	Also propagate by division of clumps
Collards	¼	10-12	10-15	24-30	4-10		•		4-6*	65-85T	Direct seed for a fall crop.
Corn, Sweet	2	4-6	10-14	30-36	6-10			•		60-90	Make successive plantings.
Corn Salad	½	8-10	4-6	12-16	7-10		•			45-55	Tolerant of cold weather.
Cress, Garden	¼	10-12	2-3	12-16	4-10		•			25-45	Seeds sensitive to light.
Cucumber	1	3-5	12	48-72	6-10			•	4	55-65	See text about training.
Dandelion	½	6-10	8-10	12-16	7-14		•			70-90	
Eggplant	¼-½	8-12	18	36	7-14			•	6-9*	75-95T	
Endive	½	4-6	9-12	12-24	5-9		•		4-6	60-90	Same culture as lettuce.
Wonder Berry Garden Huckleberry	½	8-12	24-36	24-36	5-15			•	5-10	60-80	
Fennel, Florence	½	8-12	6	18-24	6-17		•			120	Plant in fall in mild winter areas.

*Transplants preferred over seed.
T Number of days from setting out transplants; all others are from seeding.

Vegetable	Depth to plant seed (inches)	Number of seed to sow per foot	Distance between plants (inches)	Distance between rows (inches)	Number of days to germination	Soil temperature for seed			Weeks needed to grow to transplant size	Days to maturity	Remarks
						Needs cool soil	Tolerates cool soil	Needs warm soil			
Garlic	1		2-4	12-18	6-10		•			90-sets	
Gourds											See text
Ground Cherry Husk Tomato	½	6	24	36	6-13			•	6*	90-100T	Treat same as tomatoes.
Horseradish	Div.		10-18	24			•			6-8 mos.	Use root division 2-8″ long.
Jerusalem Artichoke	Tubers 4		15-24	30-60			•			100-105	
Jicama	¼	—	6-8	—	7			•	2	4-8 mos.	Seeds costly, start indoors in peat pots.
Kale	½	8-12	8-12	18-24	3-10		•		4-6	55-80	Direct seed for fall crop.
Kohlrabi	½	8-12	3-4	18-24	3-10		•		4-6	60-70	
Leeks	½-1	8-12	2-4	12-18	7-12		•		10-12	80-90T	130-150 days from seed.
Lettuce: Head	¼-½	4-8	12-14	18-24	4-10	•			3-5	55-80	Keep seed moist.
Leaf	¼-½	8-12	4-6	12-18	4-10	•			3-5	45-60	Keep seed moist.
Muskmelon	1	3-6	12	48-72	4-8			•	3-4	75-100	
Mustard	½	8-10	2-6	12-18	3-10		•			40-60	Use early to thin.
Nasturtium	½-1	4-8	4-10	18-36			•			50-60	
Okra	1	6-8	15-18	28-36	7-14			•		50-60	
Onion: sets	1-2		2-3	12-24		•				95-120	Green onions 50-60 days.
plants	2-3		2-3	12-24		•			8	95-120T	
seed	½	10-15	2-3	12-24	7-12	•				100-165	
Parsley	¼-½	10-15	3-6	12-20	14-28		•		8	85-90	
Parsnips	½	8-12	3-4	16-24	15-25		•			100-120	
Peas	2	6-7	2-3	18-30	6-15	•				65-85	
Peanut	1½	2-3	6-10	30				•		110-120	Requires warm growing season.
Peppers	¼	6-8	18-24	24-36	10-20			•	6-8	60-80T	
Potato	4	1	12	24-36	8-16		•			90-105	
Pumpkin	1-1½	2	30	72-120	6-10			•		70-110	Give them room.
Purslane	½	6-8	6	12	7-14	•					
Radish	½	14-16	1-2	6-12	3-10	•				20-50	Early spring or late fall weather.
Rhubarb	Crown		24-30	36				•	1 yr.	2 yrs.	Matures 2nd season.
Rocket	¼	8-10	8-12	18-24	7-14		•				
Rutabaga	½	4-6	8-12	18-24	3-10		•			80-90	
Salsify	½	8-12	2-3	16-18		•				110-150	
Salsify, Black	½	8-12	2-3	16-18		•				110-150	
Shallot	Bulb—1		2-4	12-18			•			60-75	
Shongiku	½	15-20	2-3	10-12	5-14		•			42	Best in cool weather
Spinach	½	10-12	2-4	12-14	6-14	•				40-65	
Malabar	½	4-6	12	12	10			•		70	
New Zealand	1½	4-6	18	24	5-10			•		70-80	
Tampala	¼-½	6-10	4-6	24-30				•		21-42	Thin and use early while tender.
Squash (summer)	1	4-6	16-24	36-60	3-12			•		50-60	
Squash (winter)	1	1-2	24-48	72-120	6-10			•		85-120	
Sunflower	1	2-3	16-24	36-48	7-12			•		80-90	Space wide for large heads.
Sweet Potato	Plants		12-18	36-48				•		120	Propagate from cuttings.
Tomatillo	½	6	24	36	6-13			•	6	90-100T	
Tomato	½		18-36	36-60	6-14			•	5-7	55-90T	Early var. 55-60. Mid 65-75, Late 80-100.
Turnip	½	14-16	1-3	15-18	3-10	•				45-60	Thin early for greens.
Watermelon	1		12-16	60	3-12			•		80-100	Ice-box size mature earlier.

*Transplants preferred over seed.

T Number of days from setting out transplants; all others are from seeding.

THE VEGETABLES

Photos and detailed descriptions give you comprehensive information about the most common vegetables grown by home gardeners around the country. Here are all the cultural requirements plus pointers for avoiding common mistakes.

It's the fashion in conventional vegetable books to present the individual vegetables in some sort of order: alphabetical; by major and minor crops; by the warm-weather and cool-weather crops; or by families. We have used a primarily alphabetical arrangement. Whatever vegetable you want to look up, check first at its alphabetical position.

Where alphabetizing would make for too much repetition, we have made some groupings by culture or other similarities. Examples are "The Root Crops" and "The Cabbage Family."

One quick way to find what you want is to use the index. It will refer you to this chapter or to "Vegetables of Special Interest," or wherever the plant is mentioned.

Where the availability of a vegetable or vegetable variety is limited, the name is followed by one or more numbers; for example: 'White Beauty' (9) and 'Tom Thumb' (5, 21). These numbers correspond to a numbered list of catalog seed sources on page 104. Wherever you see a number in parentheses, it refers to these seed sources.

ASPARAGUS

Asparagus is fast-becoming one of the most popular home garden vegetables, and for some very good reasons. It is relatively easy to grow. It is an expensive vegetable to buy at the supermarket. You also can plant this perennial once and harvest early each spring for

Fresh vegetables arrayed in the kitchen. As any home gardener will tell you, home-grown vegetables aren't necessarily better than their store-bought counterparts—they just taste that way.

a dozen years or more.

Asparagus is the earliest vegetable that can be grown and harvested from the garden, and it thrives in most areas of the US. Ideal climates are cold enough in winter to freeze the soil a few inches deep. (It does not thrive in the Deep South states of Florida, Louisiana, and Alabama.)

A plot 20 feet square or a row 50 to 60 feet long will keep a family of 5 or 6 well supplied with fresh asparagus.

Preparing the soil: When planting asparagus you are building the foundation for 10 to 15 years of production; so take the time to work the soil a foot or more deep, adding in plenty of organic matter. At the same time apply 4 to 5 pounds of 5-10-10 fertilizer per 100 square feet.

Asparagus is usually started with 1-year-old crowns. This not only saves time but insures planting a vigorous, productive variety. Crowns are available from garden centers and seed companies. Select large, well-grown crowns that have many roots. Thinly rooted crowns are a common cause of weak plants. Roots must not be allowed to wither or to dry out.

Asparagus roots spread wide; so dig trenches 8 inches deep and 4 to 5 feet apart. Spread some compost or manure in the bottom of the trench and cover with an inch of garden soil.

Set the crowns 18 inches apart in the row and cover with 2 inches of soil. As the new shoots come up, gradually fill in the trench.

For high production and thick spears follow a twice-a-year feeding program. Make one application before growth starts in the spring and a second as soon as the harvest is done to encourage heavy top growth.

Don't skimp on water when the tops are developing.

To harvest, cut or snap off the spears when 6 to 8 inches high. A handy tool for cutting is the asparagus knife. "Snapping"—bending the spear over sharply until it breaks—avoids injury to other shoots below ground.

No cutting should be done the first year and only a few stalks should be cut the second year. This way the plant builds up a large reserve of root power. (Note: If 2-year-old roots are available in your area, they still require this 2-year establishment period.)

Asparagus spear

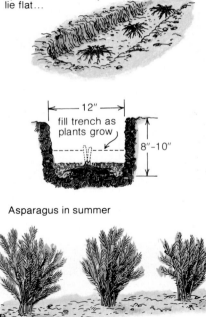

Set crowns in trench 18" apart—spread roots so they lie flat...

12"

fill trench as plants grow

8"-10"

Asparagus in summer

The third year in your garden asparagus should give you 4 full weeks of cutting. Early in the season, shoots may require cutting only every third day; but as weather warms and growth becomes faster, it may be necessary twice a day.

Varieties: Choose rust-resistant varieties such as 'Mary Washington' and 'Waltham Washington'.

How to use: Young, fresh asparagus is excellent raw. Slice thinly on the diagonal for salads, or serve it on relish trays with sauces or sour cream dips.

Nothing kills the appearance and taste of asparagus like overcooking. As Nero said, "Execute them faster than you cook asparagus." Spears should be cooked until flexible, but never soft (5 to 15 minutes, depending upon size). For small portions, use a steamer basket; for larger amounts, tie into bunches and stand upright in a coffee pot or an inverted double boiler.

Beans

Beans originated in Central America, but were well distributed in many parts of the western hemisphere before Columbus arrived. Several varieties common today were developed from beans grown by the American Indians.

Snap beans, also known as garden beans, green beans, and string beans, are all grown for their immature green pod. (The "wax" bean, however, has a yellowish, waxy pod.) Shell beans are grown for their immature green seed. Dry beans are allowed to fully mature, then are collected and stored. Many beans may be used in one or more of these forms.

Beans may be termed "bush" or "pole," depending upon growth habit. Bush beans grow 1 to 2 feet high and are usually planted in rows. Pole beans require the support of a trellis or stake. They grow more slowly than bush types but produce more beans per plant.

The many types of beans discussed here will vary in their heat requirements and length of time needed to make a crop. All except the fava bean group require warm soil to germinate and should be planted after the last frost in spring. Planting season length varies by climate. Check the charts on pages 28-33 to find the length of your growing season. Knowing this and the number of days the crop needs to mature will enable you to determine how many crops are possible, and the last planting date of the season.

Beans require various degrees of nitrogen-fixing bacteria to be present in the soil. If your soil is lacking in the bacteria, an inoculant will certainly help. If you've grown beans without any problems, you probably do not need it; but if you've never grown them and want to be sure, you can buy an appropriate legume inoculant.

Beginners' Mistakes With Beans

Planting too early: Bean seeds will not germinate in cold soil. If you need to advance the season, sow seeds indoors in peat pots and set out plants when the soil warms.

Leaving overmature pods: To get a full crop of snap beans, pick them before large seeds develop. A few old pods left on a plant will *greatly* reduce the set of new ones. Keep them picked in the young, succulent stage.

Allowing soil to dry: Lack of moisture in the soil will cause the plant to produce "pollywogs"—only the first few seeds develop and the rest of the pod shrivels to a tail.

Lack of food: Beans must make strong growth to be good size before flowering. Mix a 5-10-10 fertilizer at the rate of 3 pounds per 100 square feet into the soil before planting.

Spreading disease: To avoid the spread of disease from plant to plant, cultivate shallowly and only when leaves are dry of dew or other moisture. Harvest only when dry.

Not rotating crops: Beans are subject to diseases that survive in the soil; therefore, growing sites should be alternated each season.

FAVAS, OR BROADBEANS, HORSE BEANS, WINDSORS

Not true beans at all, these are related to another legume, vetch. They grow in cool weather unsuitable for snap beans, and will not produce in summer heat. In mild-winter areas they are planted in the fall for a spring crop. The plants grow 3 to 4½ feet high, and pods should be harvested when the seeds are half-grown.

How to use: If harvested when seeds are pea-size or smaller, favas can be treated as snap beans. Normally, however, they're allowed to reach full growth, which requires shelling and peeling before cooking. Use any recipe for limas, and serve these beans with plenty of butter. They have a sweet flavor and go particularly well with ham, pork, or chicken. Put them in soup, try them puréed, or steam them and dress with a sautéed mixture of onions, garlic, and parsley. Favas also make excellent dry beans.

GARBANZOS, CHICK PEAS, OR GRAMS

Botanically, garbanzos are neither beans nor peas, but *Cicer arietinum*, the chick pea or gram. This bush-type plant is similar to snap beans in culture but requires a longer growing season, about 100 days. Garbanzos produce one or two seeds in each puffy little pod. Pick in the green-shell stage or let them mature for dry beans.

How to use: Garbanzos can be picked green and eaten raw, but the dried, cooked form is most familiar. Whole garbanzos enhance a tossed or mixed bean salad, and may be simmered with sautéed onions and served with mushroom sauce, or added to

Yellow wax beans

soups or casseroles, particularly in combination with other legumes.

Garbanzos are a prime ingredient in the Spanish *olla podrida*, a hearty meat and vegetable stew. The Middle East has given us puréed beans with garlic, lemon juice, sesame paste, chopped red pepper, and spices. Vegetarian cooks often combine garbanzos with grains to form high-protein meat substitutes.

HEIRLOOM BEANS

Hundreds of varieties of beans have been lost to general circulation. Some have been preserved by individual gardeners or families, so are called by the general term, "heirloom beans." A variable lot, they are above all else interesting and fun for the experimenter. For more about heirloom beans, check with seed sources (60, 72).

HORTICULTURAL BEANS

These large-seeded beans are grown primarily for use in the green-shell stage, the fiber being too tough for snap beans. The mature pods are colorful, striped, and mottled, usually with red.

Harvest when the pods begin to change from green to yellow color.

Varieties: Check (15, 19, 72) for horticultural varieties.

How to use: This is the French flageolet bean, used exclusively by the better French restaurants and served wherever French gourmets gather. To savor their superb flavor—a rich, meaty taste—they must be eaten in the green-shell stage. Cook in the manner of fresh limas. Sauté in butter, add a dash of tarragon, and serve with roast leg of lamb. Flageolets also freeze and can well, and as dry beans are prized for the way they hold their shape during cooking.

LIMA BEANS

Limas need warmer soils than snap beans to germinate properly, and higher temperatures and a longer season to produce a crop. If days are extremely hot, however, pods may fail to set. If soil temperature is below 65 degrees F., pre-treating seed with both an insecticide and a fungicide is good insurance.

Harvest lima beans as soon as the pods are well filled but while still bright and fresh in appearance.

Varieties: For bush limas, consider 'Baby Fordhook', 'Fordhook', 'Fordhook 242', 'Henderson Bush,' and Jackson Wonder'. Promising new varieties are 'Bridgeton' and 'Early Thorogreen'.

The most popular pole limas are 'Carolina' (or 'Sieva') and 'King of the

Scarlet runner beans

Garden'. In Florida, look for 'Florida Butter' (or 'Calico').

How to use: To hull limas, press firmly on the pod seam with your thumb. Beans should pop out easily. Steam or simmer them (20 to 40 minutes, depending on size).

Try limas with sautéed onions and mushrooms or crumbled bacon; or with dressings of tomato sauce, warm cultured sour cream, or lemon butter and dill. Lima bean soup, simmered all day with a ham bone, makes a hearty supper. Mix with corn for that old southwestern favorite, succotash, or serve cold with red onions and parsley in a vinaigrette marinade.

SCARLET RUNNER BEANS

These beans are closely related to common beans, but are more vigorous and have larger seeds, pods, and flowers. The plant will grow rapidly to 10 and even 20 feet, forming a dense yet delicate-appearing vine with pods 6 to 12 inches long. Because of its large clusters of bright red (sometimes white) flowers, it is often grown as an ornamental.

Only pole varieties are commonly available. Culture is similar to the other beans, but give them more space.

How to use: Young pods picked when the seeds just start to develop may be cooked as green beans. Older, they are tough and strong-flavored. Seeds may be shelled and are delicious when used like limas; however, though sometimes even sold as limas, they lack that distinctive flavor. Dry, the large black-and-red beans are strong-flavored and after cooking, somewhat astringent. They remain

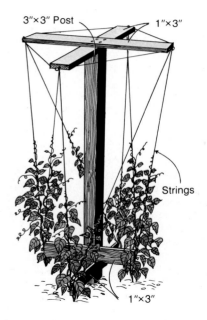
The small-space gardener will find that production is better and longer with pole beans.

popular in the Southwest, where they are often used like pinto beans in chili recipes.

SNAP BEANS

Snap beans are said to be the foolproof vegetable for the beginning gardener. They require only about 60 days of moderate temperatures to produce a crop of green pods. With such a short growth period, they can be grown throughout the United States, and in most areas can be harvested over many months from small plantings every 2 weeks. In long-season areas, snap beans may be grown virtually all year, but podding should not occur when weather is too hot or too cold.

Bush snap beans are slightly more hardy than pole kinds, and generally can be planted as much as 2 weeks earlier. They are also somewhat less susceptible to heat and drought.

Varieties: The following bush snap beans are always near the top of recommended lists: 'Bush Blue Lake', 'Commodore', 'Contender', 'Green Pod', 'Improved Tendergreen', 'Roma', 'Romano', 'Royal Burgundy', 'Tenderette', 'Tender Crop', and 'Top Crop'.

Look for 'Green Crop', recently chosen as an All-America Selection. Interesting new varieties are 'Gator Green', whose pods form just above the ground; and 'Daisy Bush Bean', whose pods form about the leaves, making finding and harvesting faster.

Four varieties of bush wax beans are consistently rated high: 'Cherokee Wax', 'Gold Crop Wax', 'Pencil Pod', and 'Rustproof Golden Wax'.

Pole snap beans produce over a longer period. The small-space gardener also may find that going vertical with a trellis, teepee, or fence of pole beans suits his garden plan better than a succession of bush beans. Poles for pole beans should be about 8 feet long.

Favored varieties of pole beans include: 'Blue Lake', 'Kentucky Wonder', and 'Romano'.

New varieties deserving a try are 'Selma Star', and 'Selma Zebra'.

In the South, look for 'McCaslan'. 'Dade' is the favorite south Florida pole bean.

The outstanding pole wax bean is 'Kentucky Wonder Wax'.

How to use: Whether bush or pole, each type of snap bean can be prepared in the same way: break off both ends of the pod, then cut into 1-inch pieces, or for French-style beans, quarter lengthwise. Boil fresh, young beans in vegetable stock for 8 to 10 minutes, then serve with butter, salt and pepper, and a few drops of lemon juice. Simmer with fried bacon squares and chopped onion; or steam briefly, then sauté in olive oil with a clove of garlic.

SOYBEANS

Soybeans were cultivated in China in 3,000 B.C. and since earliest times have been an all-important food in Manchuria, Korea, and Japan. First brought to the United States in 1804, they were used mainly as a forage crop until 1920. In 1942 the wartime demand for edible oils and fats created a boom in commercial planting for seed.

An exceptionally rich source of protein and a staple of diets throughout the world, soybeans are now being recognized as a superior home garden vegetable.

Most varieties have a narrow latitude range in which they will mature properly and produce a satisfactory crop. While short nights (long days) delay flowering, and long nights (short days) speed flowering, varieties chosen for your latitude range will mature at the same date regardless of when you plant.

Soybeans are grown similarly to snap beans (see the planting chart). However, early varieties can be planted as close as 2 feet between rows; late-maturing varieties require 3 feet.

Avoid cultivation or harvest when the plants are wet, and are thus easily bruised and broken. Wet leaves also facilitate the spread of disease. Harvest for green beans as soon as the pods are plump and the seeds nearly full size but still green. All the beans on the plant will ripen at about the same time, so you might as well pull the plant, then find a shady spot to pick the pods.

Varieties: Check seed sources (4, 5, 11, 12, 18, 19, 27, 44, 45, 60) for soybean varieties.

How to use: Most home gardeners use soybeans in the green stage. The pods are easily shelled if plunged into boiling, salted water for five minutes, cooled, then gently squeezed. Shelled soybeans take about 15 minutes to cook and are good with any garnish used for limas.

High protein content increases the soaking and cooking time for dried soybeans. A shortcut for all dried beans is to cover with water in a kettle and boil for 2 minutes, remove from heat and allow to stand 1 hour still covered with water, then cook until tender.

In dried form, soybeans can be sprouted, ground into flour, or made into soybean oil, milk, or the curd called *tofu*. They are often used as a meat substitute.

YARDLONG BEANS, OR ASPARAGUS BEANS

Catalogs list this plant as a "heavy producer of 24-inch-long beans," but actually it is a tall-growing variety of cowpea. Train this vigorous climber on wire or some type of trellis.

Harvest pods when young.

How to use: Treat as snap beans. Break yardlongs into 1 or 2-inch pieces and cook quickly to preserve their delicate, asparagus-like flavor. Simmer 3 to 5 minutes and serve with butter and seasonings; or pan fry and serve with rice.

Beets
See the Root Crops.

Broccoli
See the Cabbage Family.

Brussels sprouts
See the Cabbage Family.

Cabbage Family
The eight vegetables of the cabbage or cole family—broccoli, brussels sprouts, cabbage, cauliflower, Chinese cabbage, collards, kale, and kohlrabi—are excellent home garden crops. They are grown in every climate of the U.S. in one season or another.

The cole vegetables are adapted to cool weather, growing best when temperatures are between 65 and 80 degrees F. Planting should be timed for harvest also during cool weather. In cold-winter areas, plant for summer and early fall harvest. In the South plant for harvest in late fall or winter. In mild climates plant for late spring or fall harvest.

All cole crops are frost-hardy, most tolerating temperatures into the low 20's F. (Chinese cabbage and cauliflower are the least hardy.)

All of the cole crops will grow well in reasonably fertile, well-drained soils. If a soil test reveals overly acid or alkaline conditions, raise pH with lime or lower it with sulfur into the 6.5 to 7.5 range. (Check the label for how much to use.) A pH within this range discourages clubroot disease and permits maximum availability of soil nutrients.

Work into the soil a pre-plant fertilizer such as 10-10-10 at the rate of 2½ to 4 pounds per 100 square feet. Soils known to be fertile may need only 1 to 2 pounds. Where rains are heavy or soil is sandy, nitrogen is quickly leached. In these cases, a side dressing of about 2 pounds of straight nitrogen fertilizer per 20 feet of row will likely be necessary.

Planting: All of the coles can be direct-seeded or transplanted. Direct-seeding is recommended most often for the more sensitive Chinese cabbage and kohlrabi. (If you must transplant these, use peat pots or similar plantable containers.)

Space seeds about an inch apart in rows and cover with ½ to 1 inch of soil. Sow about 2 weeks earlier than you would set out transplants, to have the same maturity date.

Transplants should be 1 to 1½

Broccoli

months old and have 4 to 5 true leaves when set into the garden. Transplants frequently have crooked stems, and these should be planted as deeply as up to the first leaves. This will insure a sturdy plant that will not tend to flop over when full-sized.

"Bolting" is the rapid formation of a seed stalk, without the previous formation of the "head" or other harvestable product. It happens typically to the biennials: cabbage, brussels sprouts, and kohlrabi. These will bolt if young plants with several true leaves are exposed to temperatures below 50 degrees F. for 2 to 3 weeks. Large transplants exposed to low winter temperatures will flower in spring, rather than make a crop. Bolting will be less a problem if transplants with stems the thickness of a lead pencil are used.

BROCCOLI

Probably the most popular cole crop with home gardeners, broccoli is highly productive; unless a considerable amount is to be preserved, 6 to 12 plants will be adequate for most families.

In most respects, broccoli culture is the same as cabbage. Most varieties will need 60 to 85 days to mature from transplants to harvest; direct-seeded plants will need another 2 weeks.

Late spring or early summer is the best time to plant in most of the U.S. In mild-winter areas plantings are frequently made in fall for a winter harvest.

Harvest the center green bud cluster while the buds are still tight and before there is any yellow color. Broccoli will continue producing bonus side-shoots as long as the harvested shoots are not cut back to the main stem. Leave the base of the shoots and a couple of leaves to allow new shoot growth, and harvest season should last a month or more. (Note: Some of the newer varieties, such as 'Premium Crop', do not form side-shoots after the main harvest.) During hot weather, however, buds will pass quickly from prime condition.

Problems: Common problems are small plants that flower early or head poorly. Bolting—flower formation before heads are harvestable—occurs during periods of high temperatures. Planting late in spring also contributes to this problem. Premature flowering may also be caused by extended chilling of young plants, extremely early planting, transplants that are too old or too dry, or severe drought conditions.

Varieties: Five of the best broccolis are ' Cleopatra', 75 days to maturity; 'Green Comet', 55 days; 'Italian Green Sprouting' (or 'De Cicco'), 65 days; 'Premium Crop', 82 days, All-America Selection; and 'Waltham 29', 74 days.

How to use: Broccoli is easily prepared by boiling or steaming, but it is delicate and must not be overcooked. The leaves contain the greatest concentration of nutrients, so don't discard, but cook along with the flowerets and stems. Flowerets also may be sautéed.

Broccoli takes well to many garnishes, seasonings, and sauces.

BRUSSELS SPROUTS

In most ways, the culture of brussels sprouts is similar to cabbage. The most cold-tolerant of the cole family, it is a relatively long-season plant.

In most areas, set out transplants in May, June, or July for fall harvest. In mild-winter areas, plant later for winter and spring use. For further information, see the planting chart on page 48.

The sprouts mature in sequence from the bottom up. Remove leaves from beneath the lowest sprouts and those sprouts above will continue to develop.

Pinch out the growing tip when the plants are 15 to 20 inches tall to promote uniform development and maturity of the sprouts. This technique is particularly helpful where winter sets in early, but may somewhat reduce harvest. If not pinched back and you want a longer harvest, do not gather sprouts at once. When all sprouts have been gathered, the most tender leaves at the plant's top may be used as greens.

In late fall, remove all the leaves from the plant and hang it in a cool, dry cellar to enjoy fresh brussels sprouts throughout the winter.

Varieties: The varieties most used are 'Jade Cross Hybrid', 95 days; and 'Long Island Improved', 90 days. Others are 'Early Morn', 105 days; and 'Lunet', 105 days.

How to use: Boil these diminutive cabbages, then dress them with butter, lemon juice, or herb vinegar. Sprinkle nutmeg on either plain sprouts or those with sauce; or garnish with raisins, slivered almonds or chopped walnuts. Combine steamed sprouts with butter and roast chestnuts. For salad, marinate cooked

Brussels sprouts

sprouts several hours in oil and vinegar dressing, then toss with cherry tomatoes and serve.

CABBAGE

Cabbage as known today was developed gradually from leafy, non-heading relatives growing wild in various parts of Europe. There is no record of hard-heading types until 1536.

Plan for a succession of a few heads at a time. Buy transplants for the earliest planting, then sow seed directly in the garden for follow-up crops.

To produce large heads, space plants 20 inches or more in rows 36 inches apart. Smaller, normally developed heads big enough for family use can be grown spaced 12 inches apart.

Cabbage, like other cole crops, is a heavy user of nitrogen and potash. Before planting add 6 to 8 pounds of 5-10-10 fertilizer per 100 square feet and work it into the soil. Follow up in 3 to 4 weeks with a side-dressing of about 1 pound of ammonium nitrate per 100 feet of row.

Rotate planting sites of cabbage and other cole crops each year to avoid clubroot disease.

Cabbage responds very favorably to the cool, moist soil under a mulch of hay or straw. A mulch will also help control weed growth without deep hoeing. This can be important because cabbages have a relatively restricted root system that can be damaged during cultivation.

Begin harvesting when heads are firm and about the size of a softball. Cut just beneath the head, leaving some basal leaves to support the new growth of small lateral heads.

Problems: In warm weather the heads of early varieties tend to split soon after they mature. One solution is to plant at any one time only the number you can use during the 2 to 3-week maturation period. Another approach is to hold off on water or partially root-prune the plant when heads are formed. (Some gardeners simply twist the plant to break some of the roots.)

Splitting is seldom a problem with later varieties maturing during cool weather.

Bolting is a typical problem with young plants, as is the failure of overwintering plants to form heads. Read "Cabbage Family" for further information on bolting.

Cabbage Varieties

Those varieties resistant to yellows, a common disease of some northern states and much of the upper South, are indicated by (YR). All-America Selections are noted as (AAS).

Cabbage

Early to medium harvest: 'Copenhagen Market', 72 days (YR); 'Early Jersey Wakefield', 63 days (YR); 'Emerald Cross Hybrid', 63 days (AAS); 'Golden Acre', 64 days (YR); 'Stonehead', 70 days (AAS); and 'Tastie Hybrid', 68 days (YR).

Late harvest: 'Danish Ballhead', 105 days; 'Burpee's Surehead', 93 days; 'Eastern Ballhead', 95 days; 'Penn State Ballhead', 90 days; 'Safe Keeper', 95 days; 'Super Slaw', 98 days; and 'Ultra Keeper', 99 days.

Red varieties: Two of the most popular are 'Red Acre', 76 days; and 'Red Head', 85 days (YR). Try the newer 'Ruby Ball Hybrid', 68 days (AAS).

Savoy varieties: These are becomng deservedly more popular with many home gardeners. Recommended are 'Savoy Ace Hybrid' and 'Savoy King', 90 days (YR). 'Chieftain Savoy', 88 days, is the long-popular standard variety.

Midget varieties: 'Baby Head', 72 days, forms very small, very hard, thick-leaved, 2½ to 3-pound heads that hold well without splitting. Seed sources; (10, 27). 'Dwarf Mordin', 53 days (YR), makes firm, round, tender 4-inch heads (9, 21). 'Little Leaguer Cabbage', 60 days, develops round, softball-sized heads.

How to Use Cabbage

For a striking and unusual coleslaw combine both red cabbage and white with diced apples and fresh pineapple chunks.

The strong cooking odor of cabbage can be cut by simply tossing a stalk of celery into the pot. Enjoy cooked cabbage with corned beef, and in Spanish and Italian boiled dinners. Stuff the head with ground pork or veal, or make cabbage rolls of leaves filled with ground meat or rice and cheese. For a Hungarian delicacy, smother cooked, shredded cabbage in sour cream, wrap in strudel dough, and bake.

Pickled cabbage, or sauerkraut, is an international favorite—with pork and sausages in the French *choucroute garni*; in the Russian sauerkraut soup; and in the Bulgarian manner, fried, then baked with rice and cheese.

CAULIFLOWER

More restricted by climate than either cabbage or broccoli, cauliflower is less cold-tolerant and will not head properly in hot weather. Its general culture and season length, however, are about the same as for early to medium cabbage.

The approximately 2 months it needs to mature must be cool. This means planting for spring and fall crops in most cases, although winter crops are possible in mild-winter areas and summer crops in some gardens at high elevations. Planting transplants a week or two before the average date of the last frost is recommended for a spring crop.

For the most vigorous seedling growth, cauliflower is nearly always grown as a transplant. Transplants must grow rapidly; if old or stunted they will produce "buttons"—very small heads atop immature plants.

Start seeds 1 to 2 months before the outdoor planting date. When the curd (head) begins to develop, gather the leaves over it and tie them together with soft twine or plastic tape. This is called "blanching," because with light excluded, the curd will remain white and tender. The leaves of the variety 'Self Blanch' curl naturally over the head when grown in cool weather.

If weather turns hot, mist or sprinkle the plants to maintain humidity and cool. Unwrap the heads occasionally to check for hiding pests.

Varieties: The most widely grown varieties include 'Early Snowball', 60 days; 'Purple Head', 85 days; 'Snow Crown', 48 days; and 'Snow King', 50 days (AAS).

How to use: Raw cauliflower is delicious dipped into the Italian *bagna cauda*, an anchovy-garlic-butter sauce, and is also good with most popular American dips.

To serve cauliflower hot, steam and cover with a nutmeg-spiced cheese sauce in the Scandinavian manner; or bake the parboiled flowerets with melted butter and grated sharp cheese. Cooked cauliflower goes well in many quiches and casseroles.

CHINESE CABBAGE, OR CELERY CABBAGE

The name Chinese cabbage covers a number of "greens" quite different in character. All are cool-weather crops and bolt to seed in the long days of late spring and summer. Grow this vegetable as a fall and early winter crop.

Sow seeds thinly and later thin to stand 18 inches apart in rows 24 to 30 inches apart. If you must transplant, start in a peat pot or similar plantable container.

Plant from early August into Sep-

Chinese cabbage

tember. Chinese cabbage requires 75 to 85 days from seed to harvest. If frost hits before the heads form, you will still get a good crop of greens.

Varieties: One of the best is 'Michihli', 75 days, an improved version of 'Chihli'.

Others are 'Burpee Hybrid', a heading type, 75 days; 'Crispy Choy', a loose-leaf type, 45 days; and 'Early Hybrid G', a slow-bolting type, 50 to 60 days.

How to use: As its name suggests, this vegetable has a flavor between celery and cabbage. It can be used like the larger Western cabbage, but is traditional in Oriental soups, sukiyakis, and stir-fry medleys. Its delicate flavor goes well with fowl. Butter-steam the leaves to accompany duck, or, for cabbage rolls, stuff with a mixture of minced chicken and pork.

Shred fresh Chinese cabbage and add fresh pineapple for slaw.

COLLARDS

Like kale, this perennial is one of the oldest members of the cabbage family. Vegetable historians claim that it has been used for food for more than 4,000 years and cultivated in its present form for 2,000 years. Europeans described it during the 1st, 3rd, and 4th centuries, and Colonial American gardeners wrote about it in 1669.

Collards have been a southern favorite for generations. Unlike their close relative, kale, collards withstand considerable heat; yet they still tolerate cold better than cabbage.

A self-blanching cauliflower

Collards are a member of the cabbage family with a long and distinguished history

Kohlrabi

been raised as long as man has raised any vegetable. Grown for its leaves, no other plant is as well-adapted to fall sowing throughout a wide area of the U.S., or in areas of moderately severe winters. Kale is extremely hardy, but does not tolerate heat quite as well as collards.

Kale grows best in the cool of fall. The flavor is improved by a frost. Transplants may be used for early spring planting, but direct-seeding is best in fall. Since kale thinnings are good eating, scatter seeds in a 4-inch band and thin finally to 8 to 12 inches in rows 18 to 24 inches apart. In other respects, culture is essentially the same as for cabbage.

Varieties: 'Dwarf Blue Curled Scotch Vates', 55 days, is the most widely planted kale.

How to use: Always remove kale's tough stems and midribs before cooking. Chop the leaves, then proceed with any cooking method you would use for spinach.

Kale

KOHLRABI

A relatively recent development from wild cabbage in northern Europe, kohlrabi was not known 500 years ago and was not noted in the U.S. until about 1800.

This unusual but little-known vegetable deserves to be grown and appreciated more. It looks like a turnip growing above ground and sprouting long stems and leaves. (In fact, it is often described as a turnip growing on a cabbage root). The flavor is similar to both turnip and cabbage, but milder and sweeter than either.

In most regards, kohlrabi is similar to cabbage in culture; but with a shorter season, it should be started in early spring so that most growth is complete before the full heat of summer. (Kohlrabi is, however, more tol-

Planting in spring and again in fall will produce a supply of greens almost every month in the year in all but the coldest areas of the South. (A light freeze sweetens flavor.)

Collards have three different planting methods, as follows:

(1) In spring, sow seed or set out plants to stand 10 to 15 inches apart;

(2) Plant close, 5 to 7 inches apart, to dwarf and make bunchy plants for harvesting leaves as needed;

(3) In summer, sow seed thinly and let seedlings grow until large enough for greens, then harvest seedlings to give 10 to 15-inch spacing.

Collards have the same fertilizer and water requirements as cabbage.

Harvesting: Successive plantings are not necessary for a continuous supply. Collards are one of the most productive of all vegetables, particularly in southern gardens. Harvest seedlings or entire plants, or gradually pick leaves from the bottom up after plants are about 1 foot high.

Varieties: Best are 'Georgia', 70 to 80 days; and 'Vates', 75 days.

How to use: Collard greens are most often served boiled with salt pork or hog jowls. The resulting juice is called "pot likker" and is eaten with hot cornbread. Collards also traditionally accompany pan-fried fish and are common in southern-style pea soup.

For variety, cook and drain collards, top with grated sharp cheese, and bake until the cheese is bubbly. Make a hot, "wilted" salad by pouring bacon drippings over shredded young collards mixed with chopped green onions.

KALE

Kale is the closest relative of the wild cabbage from which all of the cole crops have been derived; and has

erant of heat and drying winds than other cole crops.) Start in midsummer to harvest in fall. In mild-winter areas, plant for winter and spring use.

Kohlrabi is usually seeded directly into the garden. Keep plants about 4 to 6 inches apart in rows 1 to 2 feet apart.

Harvest while the plant is 2 to 3 inches in diameter, as it becomes increasingly tough with larger size.

Varieties: Recommended kohlrabi varieties include 'Early Purple Vienna', 'Early White Vienna', 'Grand Duke', an All-America Selection; and 'Prima Hybrid'.

How to use: Use the root, discarding the stems and leaves. Young kohlrabi makes a delicious chilled salad. Or cut it into strips for the relish tray. Try marinating it in a dressing of mayonnaise and sour cream seasoned with mustard, dill seed, and lemon juice.

When small and tender, kohlrabi is good steamed, without peeling. (As it matures, it's best to strip off the tough, fibrous skin.) Dice or quarter, boil in a small amount of water, then serve with a creamy cheese sauce lightly seasoned with nutmeg.

Cabbage
See the Cabbage Family.

Cantaloupe and Muskmelon
See Melons.

Carrots
See the Root Crops.

Casabas
See Melons

CAULIFLOWER
See the Cabbage Family.

CELERY AND CELERIAC
These members of the parsley family probably originated near the Mediterranean, but wild forms barely resembling our garden kinds grow in low-lying wet places throughout Europe and southern Asia. The French apparently were the first to use celery as a food, around 1600. Earlier it found some use as a medicine. The first recorded commercial production of celery in this country was in Kalamazoo, Michigan, in 1874.

CELERY
Celery demands more time and attention than most garden vegetables. If you don't start from transplants, sow seed 2 to 4 months before your spring planting time. Seed should germinate in 2 or 3 weeks.

Sow the very small seed ⅛ inch deep and keep moist by covering the flats or pots with moist burlap. Transplant carefully, and provide generous shade and moisture to the new plants.

Celery grows naturally in wet, almost boggy locations, so the water supply must be plentiful and continuous.

Use plenty of 5-10-10 fertilizer, since celery is a long, heavy, feeder.

Although blanching usually is not necessary with modern varieties, it may make the stalks more tender. White stalks are less common in markets now and you may want to grow your own. Wrapping with paper or shading with boards will blanch the stalks.

Varieties: Use slow-bolting varieties for early spring planting: 'Summer Pascal', 115 days; and 'Golden Self Blanching', 115 days.

For late spring or summer plant-

Celery

ing use: 'Utah 52-70', 125 days; or 'Giant Pascal', 125 days.

How to use: Celery is well known as an hors d'oeuvre stuffed with cheese or dipped into hot or cold dips. Add to salads or use chopped to enhance spreads. The leafy tops, chopped fine, go well in soups and salads, and leaves also can be dried, powdered, and used as a seasoning.

By itself, celery makes a delicious creamed soup. As a hot vegetable, it can be boiled, braised, fried, or baked. Try stewing with tomatoes, shallots, and basil; or serve hot adorned with anchovy fillets and wine vinegar.

CELERIAC
A form of celery grown for its swollen root, this plant is smaller and its foliage a very dark green.

Grow celeriac in the same way as celery; it is just as demanding of high fertilizing and a continuous supply of water.

Harvest the root once it is 2 inches or more in diameter.

Celeriac

How to use: The taste of celeriac has been described as, "celery flavored with English walnuts." Though it can be shredded and served raw in salads, it is better cooked, and is good in soups, stews, and Oriental dishes. Try it steamed and served with butter or a cream sauce; or parboil, slice, and bread it, and fry it in butter.

The faint bitterness of celeriac can be removed by blanching in salted water and lemon juice just prior to preparation.

CHARD
Chard, or Swiss chard, is a kind of beet that makes edible leaves and stalks instead of roots. Considered the beet of the ancients, it was popular long before Roman times.

'Rhubarb Chard'

Chard's greatest virtue is its ability to take high summer temperatures in stride while spinach and lettuce bolt to seed.

Plant chard the same time as beets: fall to early spring in mild-climate areas; and spring to midsummer in the north. Sow seeds in rows 18 to 24 inches apart, and thin to 4 to 8 inches apart. Thinnings can be used for greens.

The large, crinkly leaves and fleshy stalks can be cut as the plant grows, so that one planting can be harvested over many months. Even if the entire plant is cut off an inch or two above the crowns, new leaves will come.

Varieties: 'Fordhook Giant', 60 days, develops very broad, thick, white stalks and thick, crinkly, dark green leaves;

'Lucullus', 60 days, has light green leaves and broad white stalks;

'Rhubarb Chard', 60 days, has green leaves and red stalks.

How to use: Chard stalks can be cooked like celery, leaves like spinach; if cooked together, the stalks should be given a 5-minute head start for equal tenderness.

Cut the thick stalks into 2 or 3-inch lengths and simmer in boiling salted water until tender. Serve hot with butter and a touch of wine vinegar, or chilled with vinaigrette.

CHINESE CABBAGE OR CELERY CABBAGE
See the Cabbage Family.

CHIVES
See the Onion Family.

COLLARDS
See the Cabbage Family.

CORN

"Corn" as a word has had many meanings. Originally it meant any hard particle of grain, sand, or salt. (Our "corned" beef earned its name because it was cured with salt.) Both wheat and barley were called corn in the Old World. Maize, the main cereal of the New World, was first known as "Indian corn" and later, just "corn."

Corn supported the early civilizations of the Americas. Fossils show that corn was grown in North America over 4,000 years ago. It was after the discovery of America that corn spread rapidly throughout the Old World.

Dent corn is considered "normal" corn. It's not sweet at maturity and has an indentation, or dent, atop each kernel. Most home gardeners grow *sweet* corn.

E. V. Wann, a research geneticist with the U.S. Vegetable Laboratory, explains the difference:

"Sweet corn differs from the other corn types by its ability to produce and retain more sugar in the kernels. This characteristic is controlled by a single recessive gene called *sugary-1*. Other distinguishing characteristics include tender kernels at edible maturity, refined flavors, a tendency to produce suckers at the base of the plant, and wrinkled seeds when dried. In recent years a new sweet corn has become available that is even sweeter than standard sweet corn, and they retain their sugar longer, for three or four days."

For a continuous supply of sweet corn through summer and into the fall, plant early, midseason, and late varieties, or make successive plantings every 2 to 3 weeks.

Keep in mind that days to maturity is a relative figure, varying by the total amount of heat the corn receives. Corn does not really start growing until the weather warms. Varieties listed as 65 days may take 80 or 90 when planted early, but may come close to the 65 days if they are planted a month later.

When planting for a succession of harvests, the effect of a cool spring should be considered. Rather than plant every 2 weeks, make the second planting when the first planting is knee-high.

Plant seed 2 inches deep, 4 to 6 seeds per foot, in rows 30 to 36 inches apart. Thin to 10 to 14 inches between plants. These are the familiar directions. You *can* crowd back to 12 inches or even the 10-inch minimum; but planting any closer risks a crop of nubbins.

Spacing: A number of the gardening specialists on our panel discussed spacing 10 to 14 inches in the row and 30 to 36 inches between rows. Here are some of their statements:

"More home garden corn plantings are *ruined* by overcrowding than any other factor."

"When seedlings are up, everyone hates to pull them out and throw them away. Actually, too many seedlings in a row act just like weeds."

"The small-growing early varieties might be spaced 8 inches apart in rows 30 inches apart, but with the later varieties such as 'Golden Cross Bantam', I like to get them spaced out to a good 12 to 15 inches for good ear production."

"If you overspace corn you generally are compensated by more usable ears and some sucker production."

How much fertilizer and when?: The consensus of our panel was:

"At planting time, fertilize in bands on both sides of seed row, 2 inches from seed in the furrow and an inch deeper than seed level. Use 3 pounds of 5-10-10 (in each band) per 100 feet of row. When the corn reaches 8 inches high, side-dress with the same amount. Repeat again when the corn reaches 18 inches (knee-high)."

Watering: It's water all the way for corn. Here's what the specialists said:

"The water need is greater from tasseling time to picking time."

"Sweet corn makes very rapid growth during the time of maturing the crop. No check in watering should occur."

"In very hot and dry weather, roll-ing of the leaves may occur in midday even when soil moisture is adequate. Plants will transpire water faster than roots can absorb it. But if leaves roll (edges turning downward), check the soil for adequate moisture."

Harvesting: As a general guide, corn is ready to harvest 3 weeks after the silk first appears. But this will vary depending upon the weather during that time. The silk will become dry and brown when the ears are near perfect ripeness. Probably the only sure way to tell is to open the husk of a likely ear and press a kernel. If it spurts milky juice, the ripeness is just right.

Varieties differ, but most will produce two ears per plant, the top ear ripening a day or two ahead of the lower one.

Harvest by breaking the ear from the stalk. Hold the ear at its base and bend downwards, twisting at the same time. The idea is to break the ear off close to its base without damaging either it or the main stalk.

Do's and Don'ts: Corn is wind-pollinated. Plant in short blocks of 3 to 4 rows rather than a single long row of the same number of plants.

Don't interplant the different types of corn. Pollen from dent corn will make sweet-corn kernels starchy and less sweet. Standard sweet varieties will reduce the quality of the extra-sweet kinds. If you must have more than one kind, wait about 4 weeks between plantings to insure that the plants will not be pollinating at the same time. Or if possible, keep dissimilar corn types 400 or more yards apart.

Don't worry about suckers. They don't take any strength from the main stalk, and removing them may actually reduce yield.

Do look out for the corn earworm. Its eggs hatch on the silk of the developing ear and larvae burrow into and feed on kernels. Ears with tight husks and good tip covers are somewhat more resistant to corn earworm damage. They do not prevent the entrance of the worms but likely will lessen the damage. Frequently, some control measure for corn earworms is necessary. See page 18.

Tight husks have an advantage over another pest. One corn-grower reports: "We have a terrible time with birds eating the kernels on the tip of the ear. The damage is worse on ears with a loose husk. And such damage to the ornamental Indian corn makes it worthless. We solved the problem by slipping a paper bag over each ear after it's pollinated."

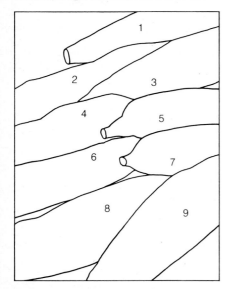

1. 'Purple Husk', 2. An inferior early variety, 3. 'Silver Queen', 4. Indian corn, 5. Indian corn, 6. Yellow hybrid, 7. Popcorn, 8. 'Country Gentleman', 9. Indian corn.

Corn Varieties

Since there are hundreds of corn varieties, some confusion over what to plant is difficult to avoid, especially if you're just getting started. There are white, yellow, bicolor, and super-sweet kinds of sweet corn. Variety names usually indicate the type. They vary, of course, in the time needed to mature. These are the results of our variety investigations.

The most popular: 'Silver Queen' is a white-kerneled extra-sweet corn and by far the most widely favored. Sweet (almost too sweet for some), with great flavor and tenderness, it has 8 to 9-inch ears and needs 92 days to mature. Plants grow to 8 feet and are resistant to bacterial wilt, a serious corn disease.

—'Golden Queen' has similar qualities but yellow kernels.

Bicolors: Kernels of these are both yellow and white, and an increasing number are becoming available. Look for:

—'Sugar and Gold', an early variety, 10 rows of kernels on 6½-inch ears, compact plants, 67 days;

—'Butter and Sugar', 12 to 14 rows of kernels on 7½-inch ears, plants 5 to 6 feet tall, 70 days;

—'Honey and Cream', 7-inch ears, for fresh use, roasting or canning, 78 days;

'Bi-Queen', 14 to 16 rows of kernels on 8½-inch ears, very similar to 'Silver Queen' in growth habit, ears, and maturity, vigorous and disease resistant, 92 days.

Old favorites: These are the widely available, long-time favorites of many gardeners:

—'Country Gentleman', large white kernels not in rows (called "shoe peg"), two or three 7-inch ears on a stalk 7 feet tall, for fresh use, canning or freezing, 93 days;

—'Earliking', yellow kernels in 10 to 12 rows on 7 to 7½-inch ears, 66 days;

—'Golden Beauty', yellow kernels in 12 to 14-inch rows on 7-inch ears, 73 days, an All-America Selection;

—'Golden Cross Bantam', the standard by which all yellow sweet corns have been measured and still a favored main-crop variety, 10 to 14 rows of kernels on 7½ to 8-inch ears, plants 6 to 7 feet tall, for fresh use, canning, freezing, 85 days;

—'Iochief', yellow sweet corn, 14 to 18 rows of kernels on 9-inch ears, usually 2 ears per 6½-foot stalk, 89 days (AAS);

—'Merita', bright yellow kernels, very productive, 16 to 20 rows of kernels on 8 to 9-inch ears, 84 days;

—'Seneca Chief', standard late or main-season yellow, 12 to 16 rows of kernels on slender 8-inch ears, sweet and tender, one of the most popular for home gardens, 82 days.

Southern varieties: Most of the preceding varieties can be grown in the South, but these long-established preferences and traditions are worth noting here:

—'Aunt Mary's Hybrid', an early white sweet corn, very sweet, stalks and ears relatively small, 70 days;

—'Hickory King', a long-season white corn and the best variety to plant for hominy, 9-inch ears, also for roasting or graining, 110 days.

—'Stowell's Evergreen Hybrid', improved yield and more vigor than the standard 'Stowell's Evergreen', 14 to 18 rows of white kernels on 7½ to 8½-inch ears, stalks 8 to 10 feet tall, 90 days;

—'Trucker's Favorite', the all-time favorite roasting white corn, 9-inch ears, stalks 7 feet high, 85 days.

Newcomers worth trying: Note the special qualities for each of these.

—'Candystick II', long and thin yellow sweet corn, small cob, good for freezing, 8 to 12 rows of kernels on 12-inch ears, 90 days;

—'Early Sunglow', very fast-maturing yellow variety, small plants about 4½ feet tall producing 6-inch ears, 12 rows of kernels per ear, 62 days;

—'Kandy Korn', yellow extra-sweet corn identified by faint red stripes and dark red tips on husks, very tender and keeps both flavor and tenderness for extended time, carries "EH" factor so does not require isolation from other sweet corns, 8-inch ears, 85 days;

—'Platinum Lady', high quality, early white sweet corn, 14 rows of small kernels on 8-inch ears.

How to Use Corn

Nothing equals the flavor of a fresh ear of corn cooked only until heated through—and "fresh" means corn brought directly from the garden to the cooking pot. The sooner the ears are used, the sweeter they'll be: even 24 hours after picking will produce a substantial loss in flavor and texture. If you must store corn, wrap it un-husked in damp paper towels and place it in the refrigerator. Then husk just before cooking, using a stiff vegetable brush to help remove the silk.

Corn-on-the-cob is usually steamed or boiled. If boiling it, add a tablespoon of sugar to the water to bring out its natural sweetness; but never add salt, which toughens the kernels. Cook until just tender, 3 to 5 minutes, then serve with plenty of butter, salt and pepper. Ears are also delicious roasted either in the oven or over hot coals. Husk them, coat with butter, wrap in aluminum foil, and roast about 40 minutes in the oven, 15 on the grill. To give this old favorite a different twist, serve the ears with melted butter seasoned with soy sauce, mixed herbs, curry powder, Worcestershire sauce, or chopped chives.

For another succulent treat, steam fresh kernels in milk or light cream. Or make corn chowder, scalloped corn, succotash (lima beans and corn), or a corn soufflé. Corn fritters with hot syrup, and sausages, make a delicious breakfast.

COWPEAS, OR BLACK-EYED PEAS
See Peas.

CRENSHAW
See Melons.

Immature cucumber with blossom

CUCUMBERS

The cucumber most likely originated in India. Vegetable historians say that it was introduced into China during the second century before Christ. The French, in 1535, found the Indians growing it in what is now Montreal, and De Soto found it being grown in Florida in 1539.

Because of its short growing season, 55 to 65 days from seed to picking size, the cucumber can find the warm slice of climate that it needs in almost every garden. But being a warm weather plant and very sensitive to frost, it should be direct-sown only after the soil is thoroughly warmed in spring and air temperatures are 65 to 70 degrees F.

Cucumbers respond to generous amounts of organic matter in the soil. For special treatment, dig the planting furrow 2 feet deep and fill the first foot or so with manure mixed with peat moss, compost, sawdust or other organic material. Fill the rest of the furrow with soil, peat moss, and 5-10-10 fertilizer at the rate of 2 pounds to 50 feet of row.

Sow seed 1 inch deep, 3 to 5 seeds per foot of row. Thin the seedlings to about 12 inches apart. Leave 4 to 6 feet between rows.

To grow cucumbers in hills, sow 9 to 12 seeds in each hill and thin later to 4 or 5, and finally to only 2 or 3, plants per hill. Hills are spaced 2 to 3 feet apart in rows 4 to 6 feet apart. There is no specific advantage to hill planting. Planting in hills probably became popular for ease of watering young plants.

Where the growing season is short in your climate, start seed indoors 4 to 6 weeks before time to set out transplants. When setting them out, cover the plants with "hot caps" or plastic to increase temperatures and protect them from frost.

Since cucumber roots will grow to a depth of 3 feet if the soil is normal, watering should be slow and deep. If the plant is under stress from lack of moisture at any time, it just stops growing. (It will pick up again when moisture is supplied.) It is normal for leaves to wilt in the middle of the day during hot spells, but check the soil for moisture below the surface.

When space is limited: Cucumbers trained on a trellis take very little ground space and produce more attractive fruits and fewer culls. Varieties that are curved when grown on the ground, such as 'Burpee Hybrid', grow almost straight when trained on a trellis.

Also consider the midget varieties when space is limited. They can be grown on the ground, in tubs and boxes, or as hanging baskets. Three such varieties are 'Patio Pik', 'Spacemaster', and 'Bush Champion.'

If the first early flowers fail to set fruit, don't worry. The male flowers open first, then about a week later you'll see flowers with baby cucumbers at their bases. These are the female flowers.

If this delayed setting does worry you, try one of the new "gynoecious" hybrids. (See under *Varieties*.) They set with the first blossoms.

Keep all fruit picked from the vines as they reach usable size. The importance of this can't be overstressed, because even a few fruit left to mature on the plant will completely stop the set of new fruit. If you can't keep up, and want the fruit to keep coming, share the harvest with your neighbors.

Bitterness in cucumbers: Dr. D. R. Bienz of Washington State University conducted a detailed study on this subject. The following are his findings:

Most home gardeners have had the experience of slicing into a fresh, crisp, green cucumber, only to find the flesh too bitter to eat. A lack of or variation in soil moisture during the growing season has often been said to cause this bitterness. Some growers feel it is more prevalent during cool growing seasons than warm ones. Faulty fertilization, harvesting during the wrong part of the day, and peeling in the wrong direction are also thought by some to contribute to bitterness. Bitterness is generally more concentrated at the stem end, but never penetrates as deeply as the seed cavity. Usually it is just under the skin and can be peeled away. The direction of peeling has no bearing on either the amount of bitterness or the amount of flesh that has to be removed to eliminate it.

Dr. Bienz found that of the several varieties tested in his study, the old and popular 'Straight Eight' developed bitterness in 80 percent of its fruits. No bitter fruit showed up on 'Eversweet', 'Improved Long Green', 'Lemon', and 'Saticoy Hybrid'.

Cucumber Varieties

There are many types of cucumbers and varieties of each type. We read the words "slicing," "pickling," "bush," "white" or "black"-spined, and "gynoecious." Then too, there's much talk about disease resistance.

Catalogs do divide cucumbers into *slicing* and *pickling* varieties. Sometimes you'll see one labeled *dual-purpose*. Some picklers can be "picked at any age," meaning small ones for sweet pickles and larger ones for dills. It is true that all cucumbers should be picked in the immature stage, but a *slicing* variety that is just right at 8 inches long does not really have the pickling quality at the small, sweet-pickle size.

Bush cucumbers are a fairly new development. David W. Davis of the University of Minnesota writes:

"Bush cucumber varieties have been under development for a long

Large cucumbers hang gracefully from an A-frame wooden trellis. They take little more ground space and produce more attractive fruits when trained this way. You easily see when they're ready to pick.

time. Some of those available trace their parentage to a very small dwarf parent type developed many years ago by A. E. Hutchins at the University of Minnesota. In cucumber, the bush and semi-bush types have seen more use by pickle manufacturing companies for mechanical harvesting. Only recently have bush varieties been available to the home gardener. 'Patio Pik' is perhaps the best known. Others advertised as having compact vines, such as 'Mincu' and 'Tiny Dill', are actually normal vines of reduced vigor."

"White-spined" or "black-spined" has nothing to do with cucumber quality when picked. The white-spined variety merely may have more eye appeal as the spines disappear.

Burpee's catalog explains:

"Cucumbers are divided into families, the 'White Spine' and the 'Black Spine.' The spines are the miniature stickers that protrude from the warts when fruits are young. White-spine cucumbers turn creamy white when old; black-spine varieties turn yellowish orange."

Lemon cucumber

Most gardeners don't leave cucumbers on the vines long enough to see which color they turn.

"Gynoecious", a word that might slow you up, simply refers to varieties that have almost all female flowers. In the regular old cucumber, male blossoms greatly outnumber the female, or fruiting, blossoms. There is no delayed setting of fruit with the gynoecious hybrids, which set with the first blossoms. They also set closer to the base or crown of the plant.

Some gardeners are lucky enough to be able to grow cucumbers without damage from its several diseases. These are: anthracnose—particularly a problem in the Southeast; downy mildew—worst in Atlantic and Gulf states; and powdery mildew, mosaic, and scab—most serious in northern states. In most areas, scab and mosaic can seriously reduce harvests.

The best insurance against cucumber disease is built-in resistance. Vegetable breeders have developed many resistant strains. In the following list of varieties, disease resistance or tolerance is indicated as follows: Anthracnose (A), Bacterial Wilt (BW), Downy Mildew (DM), Leaf Spot (LS), Mosaic (M), Powdery Mildew (PM), and Scab (S).

Slicing cucumbers: 'Ashley', released by Clemson University, widely adapted, dark green 8-inch fruits, 65 days (DM);
—'Burpless', more digestible, may be pickled, 10 to 12-inch fruits, straight if grown on a trellis, 62 days, some disease tolerance;
—'Early Sure Crop', vigorous, widely adapted, medium green 8 to 9½-inch fruit, All-America Selection, 58 days (DM, M);
—'Gemini 7', Clemson University, gynoecious, dark green 8 to 8½-inch fruit, 60 days (A, DM, LS, M, PM, S);
—'Lemon', easily grown old-timer as burpless as any, lemon shape and size, at maturity turns lemon then golden yellow, 58 days;
—'Marketer', very uniform 8 to 9-inch fruit, 65 days (DM, M);
—'Marketmore', developed at Cornell University, one of the finest main crop slicers for northern areas, 8-inch fruits, 67 days (M, S);
—'Poinsett '76', Clemson University, widely adapted, mid to early, 7½-inch fruits, 63 days (A, DM, LS, PM);
—'Pot Luck', early dwarf vine for containers or limited space, dark green 6½ to 7-inch fruits, 58 days (M, S);
—'Saladin', European origin, gynoecious, semi-smooth, tender, non-bitter skins, bright green 5-inch fruit, 55 days (BW, M, PM);

—'Sweet Slice', 10 to 12-inch very mild fruit, straight if trellis grown, All-America Selection, 63 days, some disease resistance;
—'Triumph', an old favorite hybrid, vigorous, compact vines, firm dark green fruit, 65 days (A, DM, M);
—'Victory Hybrid', gynoecious, deep green 8-inch fruit, All-America Selection, 60 days (A, DM, M, PM, S).

Pickling cucumbers: 'Liberty', North Carolina State University, excellent home garden pickler, vigorous, good cold tolerance, All-America Selection, 56 days (DM, M, PM, S), also resistant to angular leaf spot and target spot;
—'National Pickling', selected for uniformity by the National Pickle Packers Association, early, excellent for home gardens, wide fruits are cylindrical and blunt-ended, dark green, 5½ inches, 50 days;

When you grow and preserve "picklers," the harvest extends through the year.

—'Peppi', early dwarf good for small gardens, dark green fruits, 48 days (DM PM, S);
—'Salty', gynoecious, 53 days (DM, M, PM, S);
—'Wisconsin SMR 18', vigorous and prolific vines, well-shaped fruits that are firm and brine well, warted, 54 days (M, S).

Bush cucumbers: 'Bush Champion', vines short and compact, 9 to 11-inch fruit over a long season, good for slicing, 60 days (M);

—'Spacemaster', widely adapted dwarf, dark green 7½-inch fruit, 60 days (M);
—'Patio Pik', gynoecious, small vine, good yielder, medium green 4-inch fruit (A, DM).

Greenhouse cucumbers: Greenhouse cucumbers cannot be grown outside as they are seedless and self-pollinating. (Insect pollination would form seed pods, making the fruit gourd-like.) These are the plastic-wrapped cucumbers offered in markets. Their culture is demanding, and not for the casual grower. The only type of cucumber grown in Europe, they are free of bitterness and easily digested. Many seed sources offer them.

How to Use Cucumbers

For freshness and crispness, it's important not to peel or slice a cucumber until just ready to use. (Many of the new varieties needn't be peeled at all.) If you want fancy, scalloped slices, score with a fork before cutting. An ice water soak will make slices even crisper for the relish tray.

This popular salad and pickling vegetable is also good served with sour cream and a sprinkling of parsley, or, for a Middle Eastern touch, with yogurt and mint. Dress slices with a vinegar-sugar-dill sauce for a Scandinavian accent.

Although cucumbers are most often used raw, many people find them more digestible cooked. Boil slices quickly and serve with melted butter or a light cream sauce; sauté with chopped tomatoes or onions; bread and fry till golden; or bake stuffed with meat, cheese, chopped mushrooms and bread crumbs.

Eggplant

One of the oldest references to eggplant is in a 5th-century Chinese book. A black dye was made from the plant and used by ladies of fashion to stain their teeth—which, when then polished, gleamed like metal.

Wild eggplant occurs in India and was first cultivated there. The Arabs took it to Spain, the Spaniards to the Americas, and the Persians to Africa.

The eggplants received in various European countries in the 16th and 17th centuries varied greatly in shape and color. The first known appear to have been of the class now grown as ornamentals, the fruit resembling an egg. By 1806, both the purple and white ornamentals were growing in American gardens.

Eggplants are more susceptible to low-temperature injury, especially on

Eggplants will probably need support when they bear.

cold nights, than tomatoes. Don't set plants out until daily temperatures are in the 70-degree range. Plants that fail to grow because of cool weather become hardened and stunted; once stunted, they seldom make the rapid growth necessary for quality fruit. If frost dates are unpredictable in your area and very late frosts common, use hot caps or plastic covers.

Because eggplant can take 150 days to mature from seed, most gardeners grow transplants started 6 to 9 weeks before the average last frost.

Transplant about 18 inches apart, in rows 36 inches apart. Soil should be fertile and well-drained. Apply a side-dressing of fertilizer in a month, and again in another month.

Plants heavy with fruit may require support, and look out for flea beetles and the Colorado potato beetle. (See page 17.)

A cubed eggplant

Harvest 75 to 95 days after setting out transplants. For best eating quality, pick fruits when young, at about ⅓ to ⅔ their normal mature size. Good fruit has a high gloss. One test for maturity is to push on the side of the fruit with the ball of the thumb; if the indentation does not spring back the fruit is slightly mature. If upon opening the fruit the seeds are brown, the best eating stage is past. The stem is woody, so cut it with pruning shears, leaving some on the fruit.

Eggplant grows well in containers. We have grown half a dozen varieties in 5-gallon containers without a failure. Using one of the synthetic soil mixes is good insurance against the diseases that plague eggplant in some areas. In containers the medium to small-sized fruits carried high on the plant are more interesting than the low-growing heavy-fruited kinds. Where summers are cool, give containers the hot spots around the house—in reflected heat from a south wall, for example.

Varieties: While large-fruited eggplants are the most popular market varieties, the home garden trend is towards the smaller, European kinds. Days to maturity in this list applies to transplants:

—'Black Beauty', very uniform, standard eggplant, about 4 large purple fruits per 18-inch plant, slightly different strains available, 70 to 80 days;

—'Black Bell', oval fruit, very dark colored and glossy, harvestable at about 6 inches, productive 28-inch plants, 60 days;

—'Dusky', one of the earliest and most productive, thin and cylindrical

jet-black fruit, harvestable at about 5 inches, 56 days;

—'Ichiban', Oriental type, 12-inch narrow fruit, productive 36-inch plants, 60 days.

How to use: This amazingly versatile vegetable can be steamed, baked, fried, boiled or sautéed, breaded, stuffed, or sauced. Combining well with cheese, tomatoes, onions, garlic, herbs, and meats, it is the delight of cooks in many countries.

Because of its spongy texture, eggplant has a tendency to become watery. You can eliminate excess moisture by slicing, salting, and draining before use. Or stake the slices, cover with a heavily weighted plate, and let stand until moisture is squeezed out. As delicious as eggplant is fried, it also tends to soak up oil, so if you're counting calories, it's better to bake than fry or sauté.

The Greek national dish, *moussaka*, combines chopped meat and layers of eggplant, topped with a bechamel sauce and sprinkled with cheese. In India cooked chunks are served in a curry sauce. French cooks use it in *ratatouille*, a blend of garden vegetables baked and served hot as a side dish or cold as a luncheon salad. Try the Italian eggplant parmesan—slices coated in egg and bread crumbs, then baked with a topping of tomato sauce and grated cheese.

EGYPTIAN ONION
See the Onion Family.

GARLIC
See the Onion Family.

GREEN ONIONS, OR SCALLIONS
See the Onion Family.

HONEYDEWS
See Melons.

KALE
See the Cabbage Family.

KOHLRABI
See the Cabbage Family.

LETTUCE
Leaf lettuce, native to the Mediterranean and Near East, is a plant of great antiquity. More than 2,500 years ago it was cultivated in the royal gardens of the Persian Kings.

Romaine lettuce

As lettuce grows, so grows the gardener. Success with lettuce, in the full sense of the word, means not only growing a quality crop but bringing it in through many months of the year in quantities that can always be used.

If you can plan for harvesting a salad bowl combining wedges of tomatoes, slices of green peppers and cucumbers, and three kinds of lettuce, you have arrived as a vegetable grower.

Lettuce is a cool-season vegetable. Seeds are usually sown directly in early spring. If your growing season is hot and short, start heading types indoors and transplant into the garden as soon as possible, allowing the plant to mature before hot weather arrives. Succession crops are sown beginning in midsummer. Mild-winter gardeners grow spring, fall, and winter crops.

Lettuce occupies the soil for a relatively short time, but every day must be a growing one, with an adequate supply of nutrients and moisture. If the growth of a young plant is checked by lack of nutrients, it never fully recovers. Fertilize the soil before planting, with 3 to 4 pounds of 5-10-10 per 100 square feet.

To plant head lettuce, sow seed ¼ to ½ inch deep in rows 18 to 24 inches apart. Thin to 12 to 14 inches between plants. Thinnings can be transplanted for a somewhat later harvest. You can also find transplants at your garden center, or grow your own indoors.

For leaf lettuce, sow ¼ to ½ inch deep. Thin to 4 to 6 inches between plants in the first thinning, and later to 6 to 10 inches. Final spacing depends upon how large the particular variety grows.

Beginners' mistakes: Never let the plant suffer from lack of moisture. The most critical period of water need is when the heads begin to develop.

Thinning is extremely important.

If you leave 2 plants of head lettuce where only 1 should grow, you'll probably harvest 2 poor heads, or none. Thinning may be troublesome, but remember, you can use the thinnings.

The best part of open-leaf lettuce is the light green leaves in the center of a nearly mature plant. If the row is crowded, all you get is a bunch of little, bitter, outside leaves.

The Four Kinds of Lettuce

Crisphead, also known as iceberg: If there is only one lettuce in the produce display, this will be it Varieties include:
—'Fulton', similar to 'Ithaca', 80 days;
—'Great Lakes', slow to bolt, crisp serrated leaves, possibly bitter in hot weather, 82 to 90 days;
—'Imperial 44, 456, and 847', medium to small heads, adapted to spring and summer growing, 84 days;
—'Ithaca', mild, non-bolting, tip-burn resistant in all seasons, may break down in late fall weather, 72 days;

Bibb type lettuce

—'Oswego', 70 days;
—'Pennlake', best in spring, large ender heads, 70 days.

Most frequently recommended for the North and Northeast are 'Fulton', 'Great Lakes', 'Ithaca', and 'Pennlake'.

Most frequently recommended in the South are 'Great Lakes', 'Imperial 456 and 847', and 'Pennlake'. All of the 'Great Lakes' strains hold up well for fall planting.

Western favorites include 'Pennlake' in spring, 'Ithaca' in both spring and summer, and 'Great Lakes' for the fall and winter crop in mild areas.

Butterhead: These are heading types in which the leaves are loosely folded. Outer leaves may be green or brownish, inner leaves cream or butter-colored. Butterhead types are not favored commercially because they bruise and tear easily, but they're no

problem in the home garden. Varieties include:
—'Bibb', small head 3½ inches across, small dark-green leaves loosely folded, bolts in warm weather, 75 days;
—'Big Boston', medium in size, smooth broad leaves are thick and crisp, bolts easily, needs cool weather, 75 days;
—'Buttercrunch', more vigorous than 'Bibb', thick leaves, heat resistant and slow to bolt, 75 days;
—'Butter King', large and vigorous, slow to bolt, 70 days;
—'Dark Green Boston', leaves thick and substantial, loosely folded, 73 to 80 days;
—'Summer Bibb', quality of 'Bibb' but slow to bolt, 77 days;
—'Tom Thumb', a miniature with tennis-ball-size heads, 65 days (5, 21).

Northern and western recommended varieties are 'Buttercrunch', 'Dark Green Boston', and 'Summer Bibb'.

Southern standouts are 'Bibb', 'Buttercrunch', and 'Summer Bibb'.

Leaf, or bunching, lettuce: These are open in growth. Many variations are common in the outer leaves; some are frilled and crumpled, some deeply lobed. Leaf color varies from light green to red and brownish red. Seed sources for hard-to-find varieties are included here:
—'Black Seeded Simpson', light green, moderately crinkled leaves, 44 days;
—'Grand Rapids', frilled and crinkled, 45 days;
—'Oakleaf', medium-size plants, heat resistant, 40 days;
—'Prizehead' (or 'Bronze Leaf'), vigorous, mild in flavor, large, broad bronze-tinted leaves, 45 days;
—'Ruby', deserves high marks for color, 50 days;
—'Salad Bowl', tender, crinkly leaves in a broad clump, heat resistant and slow to bolt, 40 days;

'Salad bowl' with romaine lettuce

—'Slobolt', a 'Grand Rapids' type that takes more heat, 45 days.

Some significant new red varieties are:

—'Cicoria Rosa d'Treviso', an Italian red lettuce, long red leaves, crisp, adds color and unique flavor to salads, good winter lettuce for mild climates (47);

—'Deep Red', color similar to 'Prizehead' but a more intense bronze-red and richer green, leaves frilled and curly, does not take heat (13);

—'Red Salad Bowl', as attractive as delicious, bronze-red leaves, crisp, easily grown (21);

—'Salad Trim', a novel red lettuce, rich purplish-red color, sweet and crisp with romaine-like leaves (9).

Cos or Romaine: These lettuces grow more upright, to 8 or 9 inches high. Leaves are tightly folded, medium green on the outside, and greenish white inside. Varieties:

—'Dark Green Cos' and 'Paris Island Cos' are widely adapted and grown.

—The newer 'Valmaine Cos' (27) is similar to 'Paris Island Cos' but tolerant of mildew.

Favorite Lettuce Varieties

Every vegetable catalog lists an unusual, or its own favorite, variety of lettuce; for example, in source 17:

—'Crisp Mint', 65 days for early harvest, or matures in 80 days. Though not a Cos, it starts out like one and can be used similarly in all early stages of growth. As it matures it develops a rippled savoy-type leaf. It is earlier than Cos lettuces, mildew resistant and mosaic tolerant.

Other special varieties and their sources are:

—'Crispy Sweet', combines Cos growth habit and butterhead flavor, fast-growing with a long harvest season, 40 days (21);

—'Domineer', darker green, earlier strain of the 'Grand Rapids' type, 40 days (27);

—'Summerlong', reliable heading, heat tolerant, 65 days (9);

—'Sweet Midget Cos', very sweet miniature, crisp, hearts greenish-white, tender, 5-inch tall plants (9);

—'Tania', a butterhead imported from England, very uniform and flavorful, mildew resistant but not heat tolerant, 65 days (13).

How to Use Lettuce

It seems almost unthinkable to suggest that lettuce be used any other way than rinsed, chilled, patted dry, and served—either solo or in combination with other vegetables—with your favorite salad dressing. However, if your garden blesses you with an over-abundance try something different: Braise it in butter and flavor with nutmeg, as the French cook does; or make cream of lettuce soup, flavored with a dash of curry and garnished with chopped, hard-cooked egg.

Try lettuce also in wilted salads; stir-fried with mushrooms and onions; or steamed in chicken broth.

MELONS

"Melon" refers to cantaloupes, muskmelons, winter melons, and watermelons. The true cantaloupe is not grown in North America, but the term is used generally to describe the early shipping types of muskmelons. While the name muskmelon commonly refers to all types except watermelon, it may mean more specifically the melons with a musky flavor.

The winter melons—casabas, crenshaws, honeydews, and Persians—are late-maturing varieties of muskmelon.

The watermelon is a different species and requires more summer heat than muskmelons.

To grow melons, first work 5-10-10 fertilizer into soil at the rate of 4 pounds per 100 feet of row, adding generous amounts of organic matter if the soil is heavy and drains poorly. When the runners are 12 to 18 inches long, fertilize again, spreading it 8 inches away from the plants. Make a third application after the first melons are set.

Melons do need space. Give them 12 inches between plants in rows 4 to 6 feet apart. (Some "bush" varieties can be spaced much more closely.)

Vines require plenty of moisture up to full growth when growing vigorously; but hold back on watering during the ripening period.

Prescription for success:

1. Select disease-resistant melon varieties.

2. In short-season or cool-summer areas, start seeds indoors 3 to 4 weeks before the outdoor planting date, or buy transplants.

3. If you live in a short-season area, look for extra-early varieties.

4. Use black plastic mulch to raise soil temperature, conserve water, and stop weeds.

5. Protect new plants or transplants with hot caps or row covers.

Melons are a welcome treat on a hot summer day. Shown here clockwise from upper left are: honeydew, watermelon, crenshaw, and cantaloupe.

CANTALOUPES OR MUSKMELONS

The oldest supposed record of muskmelons dates back to 2,400 B.C. in Egypt. The first introduced into Europe are said to have come from Egypt to Rome in the 1st century. Columbus reported on his second voyage to the New World that he found them grown in the Galapagos from a planting two months earlier; the planting date has been established as March 29, 1494. Muskmelons were recorded in Mississippi and Alabama in 1582. and in Virginia and along the Hudson River by 1609.

Cantaloupes require from 70 to 100 days from seed to maturity. Gardeners in short-season or cool-summer areas start transplants in peat pots or other containers 3 to 4 weeks before time to plant out.

Some gardeners may manage to pick all muskmelon types at just the right time; but after trying all the "sure ways" to select the perfect melon, both in the market and the garden, we are still uncertain about its exact stage of ripeness.

A general guide is that melons shipped to markets are usually at the "full-slip" stage, meaning that the stem breaks away cleanly with slight pressure. "Vine-ripe" is when the stem breaks cleanly when you just lift the melon.

Varieties: Disease resistance in these recommended varieties is indicated as follows: Fusarium Wilt (F); Powdery Mildew (PM).

—'Banana Muskmelon', salmon flesh with faint banana flavor, banana shape, 14 to 18 inches long, 96 days to maturity;

—'Delicious 51', orange flesh, fruits 6 to 7 inches, 5 pounds, 86 days (F);

Melons need support to grow upright. Here, old pantyhose are recycled.

—'Hales Best Jumbo', close and heavy netting, grey-gold at maturity, slightly oval 6½ to 7½-inch fruit, firm flesh, popular in the Southeast, 87 days;

—'Hearts of Gold', 5 to 6-inch fruit, 3 pounds, small seed cavity, 90 days;

—'Honey Rock' (or 'Sugar Rock'), salmon flesh, 6-inch fruit, 2½ to 3 pounds, 85 days;

—'Pride of Wisconsin' (or 'Queen of Colorado'), very large-growing, deep orange flesh, 80 days (F);

—'Rocky Ford' or 'Netted Gem', prolific, favorite green-flesh type, 2 to 2½ pounds, 90 days;

—'Samson F1 Hybrid', heavy fruits with high sugar and deep flesh, All-America Selection, 90 days (F, PM);

—'Super Market Hybrid', salmon flesh, sandstone skin at maturity, 6 to 7 inches, 90 days (F, PM).

WINTER MELONS

These would be more aptly named "long, hot summer" melons. The casaba, crenshaw, honeydew, and Persian melons find their best growing conditions in the hot interior valleys of California, Arizona, and the Southwest.

The casaba is wrinkled and golden-colored when mature, with very sweet and juicy white flesh. A good keeper, casabas appear in markets for months after harvest season. Try 'Golden Beauty', 120 days (5, 6, 37).

The crenshaw has dark green skin and salmon-pink flesh and is famous for its distinctive flavor. A tender rind makes shipping a problem. Crenshaws mature in about 110 days (5, 6, 10).

The skin of the honeydew is creamy white, smooth, and hard. The flesh is lime green, with a slight golden tinge at maturity, which arrives at about 112 days (5, 15, 19).

The Persian melon is large, round, and heavily netted with thick orange flesh. It needs about 95 days in a hot, dry climate (37).

Gardeners in the north or anywhere summer is too short to accommodate these long-season melons should try the following:

—'Earlidew', a hybrid honeydew, green flesh, 5-inch fruits, ripe at full slip, 95 days (10, 19, 27, 28);

—'Earlisweet', a cantaloupe, flesh salmon-colored and thick, numerous 5-inch fruits, tolerant of fusarium wilt, 68 days (6, 28);

—'Minnesota Midget', a cantaloupe, 4-inch fruit, very compact vines requiring only 3 feet of space, 60 days (5, 9).

One of the most popular novelties

we've grown is the egg-shaped Japanese melon about the size of a lemon. It's eaten like a pear—crisp rind, white flesh, and all. Its name is 'Honey Gold No. 9', or 'Golden Crispy Hybrid' (19).

WATERMELONS

The culture of the watermelon in North Africa is believed to go back to prehistory. It is of great antiquity in the Mediterranean lands, and has been cultivated in Russia, the Near East, and Middle East for thousands of years.

The European colonists brought watermelon seeds to North America, and the plant is recorded as "abounding" in 1629 in Massachusetts.

The watermelon requires more summer heat than muskmelons. In areas not hot enough for the 25 to 30-pound watermelons, but where muskmelons can be grown, the icebox-size varieties are the best bet.

'Crimson sweet' watermelon

Picking a watermelon when it's neither too green nor too ripe is not easy. Some claim ripeness when the little "pig's-tail" curl at the point of attachment to the vine turns brown and dries up; but in some varieties it dries up 7 to 10 days before the fruit is really ripe. The sound of a thump, a ringing sound if the fruit is green, or a dull or dead one when the fruit is ripe, is also unreliable because the dull-dead sound is also the sign of overripeness.

The surest sign of ripeness in most varieties is the color of the bottom surface. As the melon matures the "ground spot" turns from light straw color to a richer yellow.

In addition, most all watermelons tend to lose the powdery or slick appearance of the top surface, becoming more dull when ripe.

Varieties: 'Crimson Sweet', for full-season areas, 15 to 25-pound fruits striped with dark green, flesh averaging 11% sugar, 97 days (F), also resistant to anthracnose;

—'Golden Midget',very early, good for northern gardens, green rind becoming golden when ripe, 8-inch fruits, very compact vine, 65 days;

—'Sugar Baby', early and productive "ice-box" melon, green stripes turning almost black when ripe, 7 to 8-inch fruits, 8 pounds, 80 days;

—'Yellow Baby', very vigorous even in short-season areas, skin thin but hard, light green with dark stripes, 7-inch fruits, 10 pounds, 86 days.

How to use: Nothing is more refreshing on a summer day than a slice of chilled, ripe melon. A ripe melon will keep in the refrigerator about a week; but always wrap a cut melon to prevent its odor from permeating other foods.

Melons alone are delicious, but they can be dressed up as well. Serve with a wedge of lemon or lime: a squeeze of juice enhances the flavor. Fill muskmelon halves with fruit, ice cream, or yogurt; or freeze melon balls or cubes in light syrup and serve in fruit compote. Peel a cantaloupe or honeydew, cut off one end, remove seeds, then fill with a fruit gelatin mixture and allow to set. Slice and serve on lettuce leaves.

Mustard

Of the many mustards, the species most frequently grown commercially is *Brassica juncea*, native to the Orient. Several varieties differing in leaf shape and texture are offered in seed packets. Another species, *B. campestris*, is listed in seed catalogs as mustard spinach or tendergreen mustard.

Mustard is a cool-weather, short-day crop and bolts to seed very early in the spring. Plant as soon as the soil can be worked in spring; and, in mild-weather areas, also in fall.

Treat the same as lettuce. Sow seed in rows 12 to 18 inches apart and thin seedlings to 4 to 8 inches apart. For tender leaves, give the plants plenty of fertilizer and water and harvest before full-grown. Mustard grows fast in fertile soil—25 to 40 days from seed to harvest.

Varieties: 'Florida Broadleaf', large, thick, green leaves with a whitish midrib, easy to clean, 50 days;

—'Fordhook Fancy', dark green leaves deeply curled, fringed and curved backward, slow to bolt, mild flavor, 40 days;

—'Southern Giant Curled', upright growth, leaves large and wide, bright green and yellow-tinged, curly edges,

Mustard

slow to bolt, mild flavor, All-America Selection, 40 days;

—'Tendergreen' or 'Spanish Mustard', produces a large rosette of dark green leaves, thick, smooth, glossy, one of the most mild in flavor, good heat resistance, 40 days.

How to use: Strong and distinctive in flavor, mustard greens are excellent served alone or mixed with milder greens such as chard. Cook them quickly in a little boiling water and dress with olive oil and white wine vinegar. For a tangy salad, combine them with beet greens, minced onion, hard-cooked eggs, and mayonnaise. Try cooking them with a mixture of sautéed bacon and onion; or simmer salt pork or ham hocks for several hours, add greens, and let cook another half hour.

Onion Family

The onion and its pungent relatives have been highly regarded since antiquity. Onions fed the sweating builders of the pyramids and the conquering troops of Alexander the Great. General Grant, in a dispatch to the War Department, wrote, "I will not move my armies without onions." The Emperor Nero earned the nickname "leek-throated" because of his frequent munching on leeks to "clear his throat."

An enthusiastic 19th century gourmet said it all for onion lovers everywhere: "Without onions there would be no gastronomic art. Banish it from the kitchen and all pleasure of eating flies with it . . . its absence reduces the rarest dainty to insipidity, and the diner to despair."

The nature of the onion is to grow tops in cool weather and form bulbs in warm weather. But the timing of the bulbing is controlled by both temperatures and day length. Onions

are so sensitive to day length that they are divided into "short-day" and "long-day" varieties. It is very important to use varieties designated for your area.

Short-day kinds are planted in the southern parts of the United States as a winter crop, started in fall. They make bulbs as days lengthen to about 12 hours in early summer. Long-day onions are grown in the northern latitudes. Most require 14 to 16 hours of daylight to form bulbs. Planted in spring as early as soil can be worked, they bulb when days are longest in summer.

'Yellow Bermuda' and 'Excel' are standard short-day varieties for the South. Because of early bulbing, they make only a small bulb in the North. (However, good-sized transplants, planted early, will make a larger bulb.) To get the very small "pearl" or pickling onions in the North, plant the short-day variety 'Eclipse' thickly in late April to May. When grown in winter in the South, it develops normal-size bulbs.

Onions are heavy feeders, so work manure and fertilizer into the soil before planting. A pound of manure per square foot and 4 to 5 pounds of 5-10-10 per 100 square feet will do the job.

Steady moisture supply is essential, particularly during bulb formation.

Start onions from seeds, transplants, or sets—the small dry onions

Green onion

available in late winter and early spring. Seed is generally the least popular, except for starting transplants. For transplants, sow indoors 12 weeks before the outdoor planting date. Sow ½ inch deep.

In the garden, make rows 1 to 2 feet apart, and thin seedlings to 2 to 3 inches apart.

Transplants are popular because of the large bulb size produced over a short time (65 days or less). Get plants from your garden center or mail-order seed dealer.

Sets are usually most reliable, though varieties available are more limited. J. S. Vandemark, Vegetable Crop Specialist at the University of Illinois, Urbana, writes:

"Select sets early when they are firm and dormant. . . . Sets are available in three colors, white, red or brown. Most gardeners prefer white sets for green onions or scallions; but the other two colors are acceptable. Divide the sets into two groups, those smaller in diameter than a dime and those larger than a dime. Use large size sets for green onions; the large size may bolt and not produce a good dry bulb. Plant the small size set for dry onion production as they will likely not bolt."

Harvest by simply pulling the onions from the ground at the time half of the tops have broken over naturally. When the tops have fully wilted, cut them off 1½ inches above the bulb. Prepare for storage by curing in an open crate or mesh bag for two weeks or more. Clean by removing dirt and outer loose dry skins, and store where air is dry and between 35°-50° F.

Varieties: For these recommended varieties we've indicated the important factor of climate adaptation as follows: North (N), South (S), and West (W). Also note that the best keepers are those with the most pungent flesh.

—'California Early Red', medium red color, flattened globe, flesh soft and mild, short keeper (W);

—'Crystal White Wax', white flesh, soft and mild, short storage, for fall planting (S);

—'Early Grano', straw-colored, shaped like a top, soft and mild, short keeper (S, W);

—'Early Yellow Globe', medium size, firm flesh, pungent, good keeper (N, W);

—'Excel', yellow, flat bulb shape, firm flesh, sweet and mild, resistant

The onion family is extensive, including scallions, leeks, garlic—and many varieties of red, white, and yellow onions.

to pink root, good keeper, fall planting (S);

—'Red Granex', red, firm flesh, sweet and mild, good keeper, fall planting (S);

—'Southport Red Globe', red, medium-size globe, good keeper, spring planting (S);

—'Southport White Globe', grown for green bunching onions or a high-quality large white onion in fall (S, N);

—'Southport Yellow Globe', yellow skin with flesh fine-grained and creamy white, medium strong, good keeper (W);

—'White Sweet Spanish', largest white, firm flesh, sweet and mild, medium keeper (N, W);

—'Yellow Bermuda', flat bulbs, soft and mild, short keeper (S, W);

—'Yellow Globe Danvers', flattened globe, medium size, firm, pungent (N, S, W);

—'Yellow Sweet Spanish', large yellow globes, sweet and mild, medium keeper, good slicer, thrips-resistant, spring planting (N, S, W).

How to use: Overcooked onions give off an unpleasant sulphurous odor, so don't cook them over high heat or too long. If you have to prepare lots of onions, drop them in boiling water for about 10 seconds, then drain and chill; the skins will slip off easily. Peeling onions under running water, of course, helps prevent tears; and to get the odor off your hands, rub them with salt or vinegar.

The onion by itself is delicious, but try small, whole ones parboiled and added to a medium cream sauce; or glaze them in honey and serve with pork. Scallop onion slices in the manner of potatoes, bake, then sprinkle with grated cheese.

Larger onions can be stuffed with sausage, beef, chicken, fish, rice, or breadcrumbs, and baked. George Washington is said to have favored mincemeat stuffing.

EGYPTIAN ONIONS

This is a very hardy and unique onion. Planted in the fall throughout the country, in early spring they may be used as green or bunching onions. By midsummer or fall they begin forming miniature bulbs at their tips, where most onions form flowers. Collect them when the tops begin to wilt and dry. Freeze them or use them fresh (11, 19, 71).

How to use: These mild-flavored onions are an excellent choice for pickling. You may also freeze them or use them fresh. The hollow stalks are ideal for stuffing like celery.

CHIVES

No vegetable gardener today would refuse to give chives room. This hardy perennial can be clipped almost continuously, and a half-dozen pots will supply enough snippings for year-round use. If not clipped, chives produce pompons of lavender flowers above their grass-like, hollow leaves.

Chives grow best in rich, moist soil in full sun but will tolerate filtered shade. The easiest way to a quick harvest is to buy plants, but you can start seeds in small pots yourself.

Look for "garlic chives." They have a mild garlic flavor and grow like regular chives, but taller. Try them in salads and stir-fry recipes (19, 65).

How to use: Chives impart a delicate onion flavor to a wide variety of dishes. Snip them into eggs, soups, sauces, cheese spreads, and dips. Sprinkle them into green salads or use to garnish cottage cheese or quiche. Spread chive butter on steaks or broiled seafood.

Chives are best used fresh, but are almost as good frozen, and still good dried. They are highly perishable, so don't add them to food until just ready to serve; and don't put them in uncooked dishes that will be stored.

Chives in bloom

GARLIC

Two types of garlic are available: the type you buy at the market, a bulb containing about 10 small cloves; and 'Elephant Garlic', which is about 6 times larger, weighing up to a pound, and has a slightly milder flavor (5, 12, 19, 21, 26).

Both types are grown in the same way: In all but the coldest areas, set out cloves in the fall an inch deep and 2 to 4 inches apart, in rows 1 to 1½ feet apart. In the coldest areas, plant in spring.

Harvest garlic when the tops fall over. Then braid them into strings or tie in bunches and hang in a cool, dry place.

How to use: Add minced garlic to salad dressings and meat sauces. In melted butter it becomes a sauce for lobster, snails, mushrooms, and many green vegetables.

Put slivers of garlic into lamb or beef roasts, or use it to season slow-cooking stews. Halve a clove to rub inside a salad bowl, or steep cloves in wine vinegar to flavor the vinegar.

Place peeled cloves in a jar or crock and cover with olive oil. Seal and store in the refrigerator. The cloves are then ready to use right from the jar, and the oil takes on a garlic pungency appropriate to many dishes. Thus preserved, cloves will keep for several months.

Setting out garlic transplants

GREEN ONIONS, OR SCALLIONS

Any variety of the standard onion can be used as a green onion if it is harvested when the bulb is small.

In addition to the bulbing type of green onions, there are several perennial "bunching" types—those that do not produce bulbs, but continue to divide at the base to form new shoots throughout the growing season. In this group are such varieties as 'Beltsville Bunching', 'Evergreen Bunching (Nebuka)', 'Hardy White Bunching', and 'Japanese White Bunching'. All are winter hardy and produce throughout the growing season. Shoots are crisp and mild early in the season, more pungent later.

The home gardener generally grows green onions from sets. Yellow sets are from the variety 'Ebenezer'. 'White Lisbon' is the most widely grown bunch onion set. The commercial green onion is always from a white variety.

The term "scallions" is used loosely to describe several kinds of onions, but is mostly applied to the non-bulbing type.

How to use: Green onion tops make a beautiful garnish for soups, egg dishes, and poultry. Enjoy whole green onions cooked: simmer in an inch of water for 3 to 4 minutes, then serve with butter and lemon juice; or chop them finely and add to your favorite quiche recipe. Try them in Oriental foods—soups, sukiyaki, and stir-fry dishes.

Many cooks prefer scallions to green onions in salads for their more delicate flavor. Scallions are delicious steamed, drained, then placed on toast and covered with melted butter or hollandaise sauce.

LEEKS

Leeks take a good 80 days to grow from transplants and 140 days from seed. When growing from seed, sow in late winter and thin to about 3 inches between plants.

Leeks do not bulb as onions do. To get long, white stems, plant them in trenches 4 to 6 inches deep and hill up the soil against the stems after the plants are fairly well grown. The most popular variety is 'Large American Flag' or 'Broad London', 130 days. The standard leek variety is 'Giant Musselburgh', maturing in about 90 days.

How to use: Leeks need to be washed thoroughly, as sandy grit burrows deep inside them. Trim the ends, slice lengthwise, and then hold under running water until clean.

In France this verseatile onion is known as "the asparagus of the poor." Steam it like asparagus or braise like celery, and serve au gratin or with a cream sauce. Try hot puréed leeks garnished with parsley, or serve leek pie as a spicy main course. For a delicious salad, chill cooked leeks and serve with hard-cooked eggs, chopped parsley, and a vinaigrette dressing.

SHALLOTS

French knights returning from the Crusades are credited with introducing shallots into Europe.

The shallot is a "multiplier" type of onion, dividing into a clump of smaller bulbs that look like small tulip bulbs. Most varieties set no seed.

Shallots are hardy and will overwinter as perennials; but for better results, lift the clusters of bulbs at the end of each growing season, and replant the smaller ones in the fall. Plant cloves 1 inch deep, 2 to 4 inches apart, 12 to 18 inches between rows (5, 12, 16, 46, 65).

Harvest when the tops die down in summer.

How to use: Shallots have a distinctive flavor somewhere between onion and garlic and are highly prized in wine cookery. Many chefs consider Bercy butter (shallots, wine, and butter) the supreme dressing for steaks and broiled meats.

It's important to mince shallots finely before sautéeing so they will cook quickly. (Never let them brown.) In most recipes you can substitute 3 to 4 shallots for 1 medium onion.

PARSNIPS
See the Root Crops.

PEAS

GARDEN PEAS, OR ENGLISH PEAS

Garden peas are known as "English peas" in the South to avoid confusing them with cowpeas.

English peas are a cool-season crop. They are grown in early spring to midsummer in cooler areas, but in fall, winter, and very early spring where it is warmer.

The low-growing varieties that do not require staking are the easiest to grow. They can be planted in rows 18 to 24 inches apart. Climbers trained on chicken wire or trellis need 3 feet between rows, but you can plant in double rows 6 inches apart on each side of the support.

Varieties: 'Alaska', one of the earliest and most popular, hardy, prolific, lower in sugar than many, uniform ripening, good for canning, plants 24 to 28 inches, 2½-inch pods with 6 to 8 peas, fusarium wilt resistant, 56 days;

—'Alderman' (or 'Tall Telephone'), 6 feet tall, needs support, 4½ to 5-inch pods with 8 to 10 large peas over a long season, 72 days;

—'Green Arrow', vigorous vines, 30 inches tall, 4-inch pods with 9 to 11 peas forming mostly at the top for easy harvest, resistant to downy mildew and fusarium wilt, 70 days;

—'Lincoln', a favored main crop pea, sweet, good fresh or frozen, 2½ feet tall, 3 to 3½-inch pods with 8 or 9 peas, an old variety productive over a long season, 65 days;

—'Little Marvel', early, very sweet, plants to 18 inches, 3-inch pods with 8 peas, 63 days;

—'Progress No. 9', early, very sweet, 20-inch plants, 4½-inch pods with 7 to 9 peas, resistant to fusarium wilt, 60 days;

—'Thomas Laxton', reliable producer, good for freezing, 3-inch pods with 6 to 7 peas, 65 days;

—'Wando', among the most heat-resistant, 30-inch plants, 3-inch pods with 7 to 8 peas, 67 days.

How to use: When small and tender, garden peas can be eaten raw in salads. Cooked, they complement any main course. Shell them just before using, and cook quickly in about an inch of boiling water.

Peas are good simply with butter, salt, and pepper, but also go well with many herbs. Try steaming them with fresh mint. Combine them in a cream sauce with pearl onions, celery, or carrots; or try them with Creole, orange, lemon, or wine sauces.

COWPEAS, OR BLACK-EYED PEAS

Middle Asia was the home of the cowpea before it migrated to Asia Minor and Africa. Then slave traders carried it to Jamaica, and in the warm climates of the West Indies, it became an important food.

Cowpeas have a more distinctive flavor than garden peas. They also require more heat—warm days and warm nights—and are damaged by the slightest frosts. Plant when the soil is warm for better germination.

Sow seed ½ to 1 inch deep, 5 to 8 seeds per foot of row, 2 to 3 feet between rows. Thin to 3 or 4 inches between plants.

Go easy with nitrogen fertilizer. A side dressing of 5-10-10 at 3 pounds per 100 feet of row after the plants are up is usually adequate.

Pick cowpeas in the green-shell stage, when the seeds are fully developed but not yet hard. Or let them ripen and store as dried peas.

Varieties: 'California Blackeye',

Blackeyed peas

These 'Sugar Snap' snap peas won an All-American Gold Medal in 1979.

large, smooth-skinned peas with good flavor, for fresh or dried use, vigorous, resistant to several pea diseases including fusarium wilt and nematodes, 75 days;

—'Mississippi Silver', a brown Crowder-type pea (seeds crowded in pods) developed by the Mississippi Agricultural Extension Service, good fresh and for canning or freezing, 80 days;

—'Pink Eye Purple Hull', the most popular southern pea, young peas are white with a small pink eye, occasionally makes two crops per plant per season, 78 days.

How to use: Cowpeas are an honored staple of southern cooking. Pick them in the green-shell stage, then shell and cook with bacon or pork. Fresh cowpeas also can be used in almost any recipe for snap beans.

To cook dried black-eyes, soak them overnight, then simmer for an hour or two with onions and bacon or salt pork. For a tasty casserole, combine cooked peas with cooked rice and bake until thoroughly heated. Tabasco is the favored seasoning for this dish.

SUGAR PEAS, OR SNOW PEAS

These are the edible-podded peas and should be picked for tenderness when very young, just as the peas start to form. If you miss that stage the pods will be too tough for eating, but you can still shell them and eat the peas.

Varieties: 'Dwarf Gray Sugar', vines to 2½ feet, can be grown without staking, light green pods 2½ to 3 inches long, 63 days;

—'Mammoth Melting Sugar', vines to 4 feet, needs support, pick when peas become just visible in the pods, resistant to fusarium wilt, 72 days;

—'Oregon Sugar Pod', productive 2-foot vines, 4-inch pods usually 2 per cluster, 68 days.

How to use: The French call the sugar or snow pea *mange-tout*—"eat it all"; and it is important in both French and Oriental cooking.

To prepare, simply snip off both tips and remove the string (if there is one). Overcooking kills both flavor and texture, so cook quickly, either stir-fried or butter-steamed. Try snow peas in soups, sukiyakis, and stir-fry combinations. Leftover snow peas are an excellent addition to salads.

SUGAR SNAP

Like the sugar or snow pea, this all-new type of pea can be eaten whole; or you can let it grow to full size, then snap and eat it like garden beans. Tall vines grow to 6 feet or more and need strong support. The pods are slightly curved, medium green, 2½ to 3 inches long, thick, and fleshy. 'Sugar Snap' takes about 70 days to mature and is best picked when the peas are large.

This very sweet pea is delicious raw—add it to salads or use with dips or on the relish tray. Steam lightly to preserve flavor. It's also good for freezing, but not for canning.

PEPPERS

Columbus, searching for a new route to the spice-laden Indies, had a different pepper in mind from those he found in the New World. The black-and-white pepper of the salt-and-pepper set comes from the berries and seeds of *Piper nigrum* (black pepper), which is in no way related to the peppers of the genus *Capsicum* (red, green, and chili peppers) that Columbus found growing in the Indian gardens in the Caribbean. His find was described upon his return to Spain as "pepper more pungent than that of the Caucasus."

Spice-hungry Europeans immediately adapted the new vegetable. Within 50 years peppers were found growing in England; in less than a century, on Austrian crownlands. They became so common in India that some botanists thought they were native there.

Peppers are classed as a hot-weather vegetable, but their heat requirements are not so high as generally supposed. Fruit set occurs in a rather limited range of night temperatures. Blossoms drop when night temperatures are much below 60 degrees F. or above 75 degrees F. Peppers thrive in areas with daytime temperatures around 75 degrees F. and nights of 62 degrees F. Daytime temperatures above 90 degrees will cause excessive blossom drop, but fruit setting will resume with the return of cooler weather. Small-fruited varieties are more tolerant of high temperatures than the larger-fruited ones.

Peppers seem to protect themselves from overloading the plant: when a full quota of fruit is underway, new blossoms drop. When some of the peppers are harvested the plant will again set fruit—if the weather is right.

The easiest way to start peppers is to buy transplants from a nursery. Varieties of sweet peppers adapted to your area are generally available. Growing your own transplants from seed is not difficult, however. Allow 7 to 10 weeks for germination and enough growth to make a good-size transplant. Peppers also can be sown in the garden in areas with long growing seasons. The shorter the season, the more reason to choose the early varieties.

Don't set out transplants until the weather has definitely warmed—at least a week or so after the last frost. When night temperatures fall below 55 degrees F. small plants will just sit, turn yellow, and become stunted.

If there is any chance of a late frost, use hot caps or similar protection.

Give the plants space to grow: Set them 24 inches apart in rows 2 to 3 feet apart. Some gardeners set out more plants than are needed and after a few weeks of growth, pull out the weaklings.

When the first blossoms open, give the plants a light application of fertilizer. Water it in well. Any stress from lack of moisture at flowering time may cause blossoms to drop. Add a mulch to conserve moisture and stop weeds.

When it's time to pick the peppers, use pruning shears or a sharp knife.

'California Wonder' bell peppers

BELL PEPPERS

The bell peppers are perhaps the most familiar peppers in the U.S. There are hot bells but most are sweet. Markets sell them green, but most will turn red, some yellow, at maturity. All mature in about 75 to 80 days from transplants.

Varieties: Resistance to tobacco mosaic virus is indicated by (TM) in the following varieties:

—'Bell Boy', plants 1½ to 2 feet tall with good leaf cover, medium-long 4-lobed fruits glossy green then deep red, All-America Selection, 75 days (TM);

—'Big Bertha', upright 2½-foot plants dark green with a thick canopy, deep green to red fruits with very thick walls, 7 by 4 inches, excellent for stuffing, 72 days (TM);

—'California Wonder', good stuffing pepper with tender flesh and delicate flavor, fruits dark green then red, 4 by 4 inches, 73 days;

—'Golden Bell Hybrid', the exception: flesh medium green maturing to deep yellow, fruit 4 by 3 inches, 68 days;

—'Keystone Resistant Giant', a popular variety for its tall sturdy plants and abundant cover with continuous production of large blocky fruit, 4-lobed fruits with thick walls, dark green then red, 75 to 80 days (TM);

—'Yolo Wonder', widely adapted market-type sweet pepper, vigorous 2-foot plants, 4-lobed fruits dark green turning red, 4 inches long, 73 days (TM).

Early Varieties: These are useful in northern gardens and several have been developed especially for that purpose. They include: 'Canape' (13, 27) 'Early Calwonder' (5, 26, 27); 'King of the North' (10, 12); 'Midway' (12, 27); 'New Ace Hybrid' (5, 27); 'Stokes Early Hybrid' (27); and 'Vindale' (27).

How to use: Most gardeners harvest bell peppers when green, but left on the plant until red they become even sweeter and more mellow. To prepare bells, remove the stems, seeds, and pith. Then slice and chop them for use in salads, soups, stews, omelets, and vegetable casseroles. If a recipe calls for peeled peppers, treat the same way as detailed under "Hot Peppers."

Chopped bell peppers sauté or butter-steam in about 5 minutes, and are especially good with onions or mushrooms. Use large, blocky bells such as 'Yolo Wonder' for stuffing with your favorite meat or vegetable mixture.

Since bells have such a short growing season and many are ready at once, dry or freeze some for later use. Cooked stuffed peppers freeze well and reheat beautifully.

HOT PEPPERS

Hot peppers are perhaps the most intensely flavored, most loved, and most confused group of vegetables. Part of the problem is the many varieties: more than 100 have been counted and they all cross-pollinate with great ease. If that is not enough, a pepper that is mild when grown in the mild conditions of a California coastal valley becomes "hot" when grown in the more stressful conditions of New Mexico.

In Mexico, all peppers are "chilis" and are commonly named by their use, such as *chile para relleno* (for stuffing), or *huachinango* (to be cooked with red snapper). Some are named for a region, such as Tabasco. Other names are derived from shape (*ancho* is broad) and color (*guero* is blond). Usually the name changes again when the chili is dried.

Dr. Roy Nakayama and his associates at New Mexico State University developed the 'New Mexico 6-4', mild-

est of the New Mexico chilis. More recently he has come up with the 'Mex Big Jim'—up to 12 inches long—a bit hotter than the 'New Mexico 6-4'. It matures early and produces more chili per acre. There may be a foot long chili relleno in our future.

Why all the interest in hot peppers? For one reason, more are produced and consumed than any other spice in the world. Dr. Nakayama expressed another reason: "One thing about chilis—once people start eating them, they get hooked."

Varieties: The following hot peppers are the most frequently recommended. Note that some are used in gardens throughout the U.S. and others are grown only in relatively specific areas. Resistance to tobacco mosaic is indicated by (TM):

—'Chile Jalapeno', widely adapted, dark green becoming red, 3½ by 1½ inches, tapered to a blunt point, very hot, 72 days;

—'College 6-4 Chile', bright red at maturity, medium thick flesh, 5½ to 6 inches by 2 inches, good fresh and for canning and drying, 78 days (6);

—'Fresno Chile', popular in the Southwest, little planted elsewhere, medium green then red, thick walls, pungent flesh, 3½ by 1-inch fruits taper to a point and are held upright on the plant, 76 days (TM);

—'Greenleaf Tabasco', especially bred for southern gardeners at Auburn University, heavy yield of red fruits excellent for sauce, 120 days (TM);

—'Hot Portugal', recommended for northern and other short-season areas, sturdy plants, heavy-yielding, green turning scarlet, 6 inches long, very hot, 64 days;

—'Hungarian Wax', light yellow then bright red at maturity, fruits 5 to 8 inches long and about 1½ inches wide, often described as "medium" hot, 67 days;

Peppers pictured here: 'Hungarian Yellow Wax', 'Serrano Chile', 'Santa Fe Grande', 'Anaheim', 'Jalapeno', 'Mercury Floral Gem'.

—'Large Cherry', medium green to red, heavy crops, fruit 1 by 1½ inches, good for pickling, hot, 78 days;

—'Long Red Cayenne', dark green then red, fruits 5 inches by ¾ inch and often curled, easily dried, very hot, 73 days;

—'Santa Fe Grande', very popular in the Southwest, yellow turning orange-red, 3½ by 1½-inch conical fruit tapers to a point, medium-thick skin, 76 days (TM);

—'Serrano Chile', highly recommended in the Southwest, green then red, small 2¼ by ½-inch fruits are slim and club-shaped, walls fairly thin, used for pickling and sauces, one of the hottest, 75 days.

How to use: Hot peppers have an honored place in international cuisine, figuring prominently in Mexican,

Peppers are as varied to the eye as they are to the taste.

Indian, African, Spanish, Portuguese, Indonesian, and Korean dishes.

Since the oil from hot peppers can irritate the eyes and skin, it's a good idea to wear rubber gloves when preparing them. You might also want to hold them under running water. Be sure to remove the stems, seeds and inner membranes (the hottest part). To peel chilis, blister them under the broiler, then slip them into a brown paper bag, twist closed, and let stand to steam and cool. Remove them one at a time and peel, starting at the stem end.

DRYING PEPPERS
Both sweet and hot peppers are dried for winter use, the dried peppers usually being called by a name different from the same fresh chili. In Mexico the dried peppers are ground in a mortar daily for a supply of chili powder. A pinch will add flavor to almost any dish, the flavor varied by adding varying amounts of the powder.

Three dried peppers important in Mexican cuisine are *ancho*, *mulato*, and *pasilla*. All are available from source (52).

JAPANESE PEPPERS
Japanese peppers are gaining popularity. An important vegetable in Japan, 'Fushimi Long Green' is a long-time favorite sweet pepper. The outstanding hot pepper in Japan is

'Yatsurusa'. Both are available from source (65).

PIMENTOS
Pimentos are the sweetest of all peppers. They are of two types: the cheese or squash; and the heart shapes. Look for 'Burpee's Early Pimento', an All-America Selection; 'Pimento Perfection', 'Pimento Trueheart', 'Tomato Pimento', and 'Yellow Cheese Pimento'. Dried, pimentos become paprika, the well known vivid red spice.

PEPPERS IN A CLASS BY THEMSELVES
Because of their special qualities, these peppers do not fit well in most classifications:
— 'Anaheim', a large-fruited cayenne, only medium hot, green fruits becoming red at full maturity, 6 to 8 inches long tapering to a point, 77 days;
— 'Cubanelle' (or 'Aconcagua'), popular yellow Italian type, red when mature, large yellow-green 6-inch fruits are smooth and tapered, a frying pepper, 68 days;
— 'Dutch Treat' (or 'Twiggy'), short plants bear 5 to 6 fruits 4 inches long, tender flesh can be eaten fresh, fruits held upright are easy to pick but susceptible to sunscald—preventable by giving them afternoon shade, All-America Selection, 70 days;
— 'Sweet Banana', like 'Anaheim' a member of the cayenne group but even more mild, thick-walled fruits are 3½ to 4½ inches long and yellow turning red.

Persian Melons
See Melons.

Potatoes
Potatoes have been grown and used in temperate climates along the Andes for at least 2,000 years; but it was not till the 16th century that Spanish explorers introduced them into Europe.

The English, French, and Germans regarded the potato mainly as a curiosity for over a century. The Irish were the first to realize its crop potential, and became economically dependent upon it by 1845, when the fungus disease *late blight* struck, eventually devastating the crop and causing widespread famine and emigration.

Potatoes became a fairly important crop in America after many Irish settlers arrived in 1718, and became very significant right after Ireland's famine.

The common "white" or "Irish" potato is now the most important vegetable in the world and the fourth most important food plant. In volume and value it is exceeded only by the grains—wheat, rice, and corn. Americans eat more potatoes than any other vegetable—about 120 pounds each per year—and for good reason: they are nutritious, economical, versatile, and easily prepared.

Potatoes require a frost-free growing season of between 90 to 120 days. The ideal climate has a relatively cool summer. Plant early potatoes *just before* the last killing frost. Sprouts will develop at low temperatures, but may be injured if exposed to frost when above ground. Usually it is safe to plant when the soil is warm enough to be worked, about 45 degrees F.

Main fall crops are usually planted 120 days before the first killing frost in fall. In short-season areas, plant as soon as possible for a late crop.

Soil preparation: Potatoes need a rich, loose, slightly acid soil. Use plenty of organic matter and add 5-10-10 or similar low-nitrogen fertilizer, about 10 pounds per 100 feet of row or as instructed on the fertilizer label.

The recommended soil pH for potatoes is 4.8 to 5.4 If the soil is not acid enough, scab disease, which causes brown corky tissue on the potato surface, may be a problem. Lime soil for potatoes *only* if a soil test shows a pH below 4.8.

How to plant: Buy certified disease-free "seed" potatoes. This is important because potatoes can host many diseases that reduce growth but are otherwise undetectable in most gardens. It also makes good sense to avoid bringing potato disease to your garden.

Good-size seed pieces increase the chances of a good yield. Cut them about 1½ inches square, making sure that each has at least one good eye, and cut a week before planting to allow cut surfaces to heal slightly. Some growers dip cut pieces in dilute bleach solution or commercial fungicide to prevent rot. You can also plant small potatoes whole and avoid the risk of rot altogether.

Set the pieces, cut side down, eye up, about 4 inches deep and 12 inches apart in rows 2 to 3 feet apart. (Ten to 12 pounds of potatoes will plant 100 feet of row and yield 1 to 2 bushels at harvest.)

Potatoes form not on roots but on stems rising from the seed. Sprouts usually appear after 2 or 3 weeks—unless they were ¼ inch or longer when planted or under less than 2 inches of soil.

If spaced right, potato foliage will shade and cool the soil as the tubers mature, preventing damaging high temperatures. To further cool the soil, mulch 6 inches deep with a loose organic material.

When plants are 5 to 6 inches high, hill up the mulch and soil around the growing stems. Potatoes exposed to light turn green, an effect associated with the naturally occurring poison, solanine. (Small amounts of green tissue can be scraped away, but excessively green potatoes should be discarded.)

Fertilizer and water: To fertilize when planting, place seed pieces in the center of a 6-inch-wide trench and work the fertilizer in at the edges with a cultivator. Do not let the fertilizer touch the seed pieces. Too much nitrogen fertilizer may cause excessive leaf growth at the expense of the tubers. Fertilize, but not too much.

Potatoes need a steady moisture supply. If soil dries out after tubers begin to form, growth stops. It starts again as soil is watered. The result of this stop-and-start growth is misshapen, knobby, split, or hollow tubers. Try to keep soil moist to a 1-foot depth through the growing season. Barring rain, that normally means one heavy watering weekly.

Harvesting: Pick "new" potatoes as soon as the tops flower. New potatoes are not a variety, but simply any potato harvested before full maturity. They are smaller and more tender, but will not store. If soil is loose, simply reach in; otherwise, gently uproot the plant to check its progress.

Potatoes headed for winter storage need to mature fully in the soil. For full-size tubers, wait until the vines yellow or die back. Store them in the dark for a week or so at 70 degrees F. to heal bruises and condition them.

Then store them at between 35 and 40 degrees F., keeping humidity high.

Beginners' mistakes: You can try growing plants from grocery store potatoes, but they frequently carry diseases and may have been treated to prevent sprouting.

Do not overfertilize before tubers are formed.

Do not ignore the best planting dates.

Do not allow tubers to receive sunlight, making them green and inedible.

Cultivate if necessary to reduce weed competition, but take care not to damage the shallow stems on which potatoes form.

Special handling: An old-time method of growing potatoes is to set seed pieces about 3 inches deep in a side trench. As stems grow, cover them with straw, leaves, pine needles, or any similar material. (If wind is a problem, anchor the straw with some soil.) Potatoes then can be picked simply by pulling back the straw.

Because a 100-foot row is needed to grow the 1 to 2 bushels most families need, potatoes are usually not considered a vegetable for small gardens. But their growth habit makes possible some interesting experiments.

Potato tubers grown in any loose material such as straw will be cleaner, better formed, and easier to harvest. One year we grew them in plastic bags, adding loose soil to fill the bag as they grew. The harvest was bountiful. They can also be grown in containers. Dr. F. W. Went, studying the effect of temperature on potato growth, planted them in straight vermiculite in 2-gallon containers. At harvest, potatoes filled ¼ of the container. (Note: Vermiculite-grown potatoes must be continuously watered with a complete nutrient solution.)

Potatoes thrive in plastic garbage bags.

Potato Varieties

There are many varieties of potatoes but only four basic types: russets, round reds, round whites, and long whites. Although most potatoes have white flesh, there are some with yellow, purple, and even bluish-tinged flesh.

In the following varieties disease resistance is indicated as follows: Late Blight (LB); Scab (S).

Early. 90 to 110 days: 'Haig', a white potato with somewhat flaky skin (S);

—'Irish Cobbler', an oblong white potato of wide adaptation, one of the earliest varieties selected, of unknown origin, susceptible to scab;

—'Norgold Russet', an oblong to long white with shallow eyes and netted skin, not as early as 'Norland', excellent for baking and boiling but does not store well (S);

—'Norland', widely adapted red potato, one of the most favored by home gardeners, very early, medium-size oblong tubers that can be used many ways (S);

—'White Rose' (or 'American Giant' or 'Wisconsin Pride'), large, long white, smooth and flattened, many eyes of medium depth, excellent for boiling or potato salad, does not store well.

Midseason. 100 to 120 days: 'Chieftain', red-skinned, smooth, attractive, seed pieces if possible should be cut 10 to 12 days before planting to allow for healing cut surfaces, good quality, many kitchen uses, some resistance to several diseases;

—'Norchip', a white potato of good eating quality (S);

—'Red LaSoda', a good red, round potato, frequently recommended in

Potatoes are easy to harvest from loose soil.

the South, one of the best adapted in Wyoming, good producer, good for boiling and potato salad, does not store well;

—'Superior', white, medium to early maturity, medium size, roundish, good yield, good cooking potato sometimes used for potato chips (S);

—'Viking', large red, excellent cooking qualities, for the best yield they should be planted close in the row to compensate for a tendency to produce a few large tubers.

Late maturing. 110 to 140 days: 'Butte', a russet-skinned hybrid very similar to its grandparent 'Russet Burbank', promising the same high kitchen quality but with more protein and vitamin C, also has shown improved yields and disease resistance;

—'Katahdin', white-skinned, widely adapted, large round to oblong tubers;

—'Kennebec', white-skinned, block-shaped, excellent eating quality, one of the best for frying and hash browns, stores moderately well (LB, S);

—'Red Pontiac', red-skinned, widely adapted, oval tubers that may become too big with abundant rainfall, fair table quality, very good storage quality;

—'Russet Burbank' (or 'Idaho Baker' or 'Idaho Russet'), the most important fall-harvested potato in the U.S. and western Canada, large tubers are long and cylindrical, a very long growing season; excellent for baking, frying, hash browns, storing (S);

—'Sebago', large elliptical white popular in northeastern and western states (LB).

How to Use Potatoes

Contrary to myth, potatoes are high in nutrition—an excellent source of protein, minerals, and Vitamin C—and relatively low in calories.

To get the most food value, leave the skin on whenever possible; if you must peel potatoes, keep parings thin. If not cooked immediately after paring, they should be covered with cold water to prevent darkening.

To prevent potatoes for a stew from becoming soggy, boil them separately, peel, and add during the last few minutes of cooking. For light, creamy, mashed potatoes, add hot milk instead of cold.

Pumpkins

See Squash, Pumpkins and Gourds.

Radishes

See the Root Crops.

Root Crops

BEETS

The original home of the beet was around the Mediterranean, where it first occurred as a leafy plant without enlarged roots. Improved types of these early beets are now grown as Swiss chard. Large-rooted beets were first noted in literature around 1550 in Germany, but there was only one variety listed in the U.S. in 1806.

Though beets prefer cool weather, they tolerate a wide range of conditions. They can be planted early, but additional plantings can be made for a long period. In very hot weather special attention to watering and mulching may be needed to get a good stand for seedlings.

Sow beets directly in rows a foot or more apart and then thin to 2 inches.

Generally, the soil and nutrient requirements for beets and turnips are the same.

Some thinning can be postponed until the extra plants are large enough for eating the greens—roots and all. Unless you use a monogerm (single-seeded) variety, each beet seedball will produce 3 to 5 seedlings in a tight clump, so some thinning should be done early.

Beginners' mistakes: The most common problems are overplanting and under-thinning.

Stringy and tough beets are the result of a lack of moisture, which may be caused by drought or by overcompetition from other beets or weeds. As with most vegetables, beets must be grown at full speed, without a letup.

Varieties: The choice of varieties for garden use is not very critical.

(Downy mildew resistance is needed in certain areas.) All varieties can serve both for roots and greens, but if greens are needed to any extent it would be better to plant chard or a variety of beet selected for that use. Sugar beets are excellent for greens. Varieties include:

—'Burpee Golden', unusual golden yellow root, good quality, may average higher in sugar, color does not bleed out in cooking, 55 days;

—'Cylindra', dark red, long cylindrical root gives many uniform slices, up to 8 inches long, 1¾ inches in diameter, 60 days;

—'Detroit Dark Red', dark color, neat globe shape, downy mildew resistant strains available, 63 days;

—'Early Blood', 68 days (9, 37, 51, 63, 69);

—'Early Wonder', semi-globe shape, 55 days;

—Egyptians: 'Crosby's Egyptian', 'Extra Early Egyptian', 'Flat Egyptian', 'Little Egypt'—dark red flesh, uniform and tasty, all early, about 55 days (11, 26, 27, 32, 46);

—'Little Ball', grows fast and stays small, serve whole, can or pickle, 56 days;

—'Long Season' (or 'Winter Keeper'), does not become as tough with increased size as most beets, tops make good greens, mature beets store well, 78 days (11, 13, 19, 27);

—'Mono-King Explorer', deep red, monogerm type, less thinning required, 50 days;

—'Pacemaker II', hybrid, globe-shaped, dark red, very tender, very vigorous and sweet, 58 days (12, 13, 16, 27);

—'Perfected Detroit Dark Red', dark red, medium-size globe, improved color, shape and uniformity compared to 'Detroit Dark Red', 58 days;

—'Ruby Queen', globe, deep red, 60 days.

Beets for greens, listed by various seedsmen, include:

—'Green Top Bunching' (6, 13, 21, 26, 27, 28, 67);

—'Lutz Green Leaf' (5, 11);

—'Sugar Beets' (9, 12, 17, 21, 47);

—'White Beet' (17, 27).

How to use: The beauty of the beet is that you can eat all of it. Cook the nutritious tops as you would other greens, then toss in a mixture of butter and breadcrumbs, or garnish with diced hard-cooked egg. Drop whole beets into boiling water; when tender, the skins will slip off easily. Cook beets with an inch of leaf stalk attached to keep them from "bleeding." A bit of lemon juice or vinegar in the water stabilizes the red pigment and adds flavor. Shred them and cook in butter; or slice them and serve hot with butter, an orange sauce, or sour cream.

Tiny beets, the thinnings from the garden, can be cooked intact—tops and roots together. Steam them in a heavy, covered pan with lemon juice, a chopped green onion, several tablespoons of oil, and your choice of seasonings.

CARROTS

Our carrots originated from forms grown in the Mediterranean. By the 13th century carrots were well established as a food in Europe and came with the first settlers to America, where Indians soon took up their culture.

Carrots are adaptable, tolerant of mismanagement, and unequaled for supplying food over a long period, using nothing more complicated for storage than the soil in which they are grown.

Most important planting times are early spring and early summer. Each planting will mature in 60 to 85 days and can be harvested over a 2 to 4-month period. Small plantings every three weeks will insure a continuous harvest.

How to grow: In the Nichols Nursery Catalog you'll find this paragraph:

"How to raise carrots without using a spade or hoe. It is simple, and here is how it is done: Build a raised bed made of 2 by 8 lumber (length optional) but width should not exceed 4 feet. Fill bed with ⅕ garden loam, ⅖ths clean sand and ⅖ths compost, rotted manure, or peat moss. For every 10 feet length of bed, spread 5 pounds of bone meal. Mix thoroughly all ingredients, then rake down into a fine seed bed. Broadcast the carrot seed, cover with ½ inch fine sifted peat moss. Water, and keep bed well moistened, but not soggy wet. Pull carrots as they are ready. July sown seed will give you carrots in the fall."

We will question, however, the inclusion of manure in the soil mix. Manures, unless very well rotted, cause roughness and branching carrots.

Carrot Varieties

Carrots differ mainly in their size and shape. The short to medium or very short types are better adapted to heavy or rough soils than the long kinds. They are also easier to dig. The tiny finger carrots are often sweetest, and are good in containers.

Long, 8 to 10 inches: 'Gold Pak', particularly deep color, 8 to 9 inches

Carrot varieties offer many shapes and sizes.

long, 76 days; 'Imperator', the standard market carrot, good quality, 8 to 9 inches long, 75 days.

Hybrid carrots are long or medium long, and normally have greater vigor and uniformity than standards. Six are: 'Pioneer' (13), 'Hipak' (13), 'Target' (13), 'Trophy' (13), 'Hybrid Sunset' (9), and 'Touché (16, 39, 40).

Medium, 6 to 8 inches: 'Danvers Half Long', almost perfectly cylindrical 6 to 7-inch roots, 1½ inches thick, 70 days; 'Nantes', favored for flavor and tenderness but tends to crack in wet fall weather, 68 days; 'Royal Chantenay', improved strain of 'Red Cored Chantenay', 6 to 7 inches long, slightly tapered, holds well in the soil, 70 days; 'Spartan Bonus', uniform, fine-grained hybrid, 6 to 7 inches, 75 days; 'Touchon', French import, practically coreless and excellent for juicing, 75 days (9, 19, 33, 36, 39, 40, 41, 46, 51).

Short, 3 to 6 inches: 'Oxheart', plump and short, recommended for heavy soils, 75 days; 'Short 'n Sweet', 3 to 4-inch bright-orange carrot, good in heavy soils (5, 17).

Finger carrots: These are ideal for containers. Their tops break off easily when pulling from heavy garden soil, so a lightweight soil mix is important. Roots are tender, sweet, and just right for eating whole or pickling.

Seven popular varieties are: 'Baby Finger Nantes' (27); 'Baby Nantes' (46); 'Bright Pak' (28); 'Lady Finger' (21); 'Little Finger' (5, 36, 39, 40, 41,

42); 'Sucram Baby' (9, 17, 19, 39); and 'Tiny Sweet' (26, 34).

European favorites: 'Amstel', a popular variety for baby carrots when thickly sown, 54 days (27, 46); 'Amsterdam Forcing', very early, very sweet, 6½ inches, 55 days (11, 27, 36, 39); and 'Belgium White', all white, very mild (19, 47).

Ball-shaped carrots: More round like a beet than long, these are excellent for container growing, serving on garnish trays, and canning in whole-pak jars.

Look for 'Golden Ball' (21); 'Gold Nugget' (9, 12); 'Parisian Ball' (39, 41, 42, 47); and 'Planet' (27).

How to Use Carrots

To make the most of this nutritious vegetable, don't pare or even scrape carrots; just scrub them well. For the relish tray, soak them in ice water for extra crispness. Shredded carrots, raisins, pineapple chunks, and mayonnaise make an excellent summer salad.

To cook, boil carrots quickly in salted water. Serve with butter and a sprinkling of parsley and chives. Added to slow-cooking stews, they improve as they take on the flavor of the meat. Cook them in a cream sauce seasoned with tarragon, nutmeg, and dill weed; or steam with mint leaves for the supreme accompaniment to lamb. Bake carrots in bread crumbs, or dip in batter and deep-fry. Try them baked in carrot bread, pies, torte, or cake.

Parsnips

White radishes

Fertilizers must be worked into the soil before planting to be quickly available to the young seedlings. Spring radishes mature in 3 to 4 weeks from seeding, so there's little time to correct mistakes.

Thin seedlings 1 or 2 inches apart very soon after emergence to reduce competition, because roots begin to expand when only 2 weeks old. Scatter seeds spaced out in a 3 or 4-inch-wide row to reduce the need for thinning. For a continuous supply of crisp roots start early with small plantings and repeat every 10 days.

Beginners' Mistakes: The most common problem among radish growers we surveyed was the cabbage maggot. Everything looks fine from the top, but at harvest, wormy radishes are discovered. For control see page 18.

PARSNIPS

Parsnips are native to the Mediterranean. They were common in Europe by the 16th century and the early colonists brought them to America.

Parsnips make 1 to 1½-foot roots in about 100 to 120 days from seed. They need loose soil, not only to grow undistorted roots, but so the gardener can harvest them intact.

Parsnips can be left in the ground all winter. In fact, they need winter cold near the freezing point to change their starch to sugar and develop the sweet, nut-like flavor they're famous for. They also may be dug in the late fall and stored in moist sand. While they can stand alternate freezing and thawing in the ground, freezing after harvest will damage them.

Varieties: The standard varieties are 'All American', 'Hollow Crown', and 'Model'.

How to use: Parsnips have a distinctive taste and a small but enthusiastic following. To prepare, steam them in their skins, then peel and slice lengthwise. If a large core has developed, cut it out. Pan glaze with butter, a touch of brown sugar, and nutmeg (for the best candied "sweet potatoes" ever!). Or brown them and dress with chopped walnuts and cream sherry.

Add parsnips to a roast the last half hour of cooking and baste frequently with the juices; or slice them into a soup or stew.

RADISHES

Give a youngster a package of radish seeds and say "go plant" and you'll have radishes. But to get crisp, mild, non-pithy radishes, you must meet the fundamentals of fertilizing and watering.

Radish Varieties

We have divided the varieties of radishes into two groups—spring and winter—even though, as the USDA points out, the term "spring" radish is misleading. They explain that spring radishes "can be grown throughout the season in cooler areas and in all but the hottest months in warmer areas."

Winter radishes are so called because they tend to flower before sizeable roots can develop when they are planted in the spring. They need the decreasing temperatures and day length of fall to discourage this flowering. The winter radishes include the Oriental, or daikon types listed below. They are slower-growing, much larger, and longer-keeping than spring types.

"Spring" radishes: These come in many shapes and sizes:

—'Burpee White', round, harvestable at ¼ to 1 inch, 25 days;

—'Champion', round, bright red, lasts in garden to 2-inch diameter without becoming pithy, 28 days;

—'Cherry Belle', round, all red, cherry-size, 22 days;

—'Crimson Giant', globe shape, 1½ inches wide, thin to 1½ inches, crisp and mild, 29 days;

—'French Breakfast', oblong, red with a white tip, 24 days;

—'Red Boy', oval, similar to 'Scarlet Globe' with extra-short tops, 22 days;

—'Scarlet Globe', olive-shaped, fast, good for forcing, 24 days;

—'Sparkler', top third red, lower third white, 25 days;

—'White Icicle', white, 5 inches, 28 days.

Winter radishes: 'Celestial', pure white, mild, 60 days (19, 22, 32, 65);

—'China Rose', long, hot, 52 days (25, 37, 40, 41, 42, 67);

—'Minowase', 50 days (17, 45, 47, 65);

—'Miyashige', 15 inches long, 2 inches wide, 60 days (19, 45, 65);

—'Sakurajima Mammoth', giant size, to 70 pounds, 70 days (10, 11, 17, 19, 25, 26, 45, 47, 65);

—'Shogoin', the standard "greens" variety, sometimes classed as rutabaga, 30 days (19, 45);

—'Takinashi', white, brittle, 12 inches long, 65 days (19, 45, 65).

How to Use Radishes

Both red and white varieties of spring radishes are most popular eaten raw, either alone or in salads. Scrub them, chill, and serve with a bowl of salt or your favorite dip.

Any radish can be cooked, but it's more common with the large winter varieties such as the daikon. Steam them with sautéed scallions, add salt and freshly ground pepper, and serve with a cream or cheese sauce.

Daikon radishes are used in Oriental dishes. Shredded daikon is the classic Japanese accompaniment to sashimi.

SALSIFY

The flavor of the salsify root earned it the names, "vegetable oyster" and "oyster plant." It is usually a biennial, growing 2 to 3 feet high. Long-stemmed purplish flower heads appear the second year.

Like carrots and parsnips, salsify grows the best roots in a deep, crumbly soil.

Plant as soon as the ground can be worked in spring. Sow seed ½ inch deep in rows 16 to 18 inches apart. When seedlings are 2 inches high, thin to 2 to 3 inches apart.

Salsify

Varieties: The standard variety of salsify is 'Sandwich Island Mammoth', 120 days. Black salsify, or *Scorzonera*, is actually a different genus, but it is used and grown similarly. It has black roots with white flesh and, in the second year, dandelion-like flowers on 2½-foot stems.

How to use: Salsify, with its distinct oyster-like taste, can be served as a main dish or to accompany any meat. To prepare, scrape the roots, cut in 2 or 3-inch lengths, and place in cold water containing a few drops of lemon juice or cider vinegar to prevent discoloration. Serve either raw or cooked.

For mock oysters, boil salsify, drain, dip in egg, then roll in flour or breadcrumbs, and sauté in butter until tender.

TURNIPS AND RUTABAGAS

Turnips originated in Western Asia and the Mediterranean in prehistoric times. Rutabagas are more recent, apparently originating in the Middle Ages from a cross of turnip and cabbage.

Though these two crops are often considered together, as here, there are distinct differences. There are white and yellow forms of each; however, most turnips are white-fleshed, must rutabagas yellow. Turnips have rough, hairy leaves, are fast-growing, and get pithy in a short time. Rutabagas have smooth, waxy leaves, emerge and develop much more slowly, are more solid, and have a long storage life. Rutabaga roots are much higher than turnips in Vitamin A and most other nutrients. The tops of both are outstanding sources of A and C.

Turnips and rutabagas prefer cool weather. Both spring and fall plant-

ings work well with turnips, as they mature in 60 days or less. Spring is best in northern areas, as early as soil can be worked. In warmer areas, fall and winter are better because the ripening period then comes at the end of the cool season.

The turnip's short season permits it to be grown at some time everywhere in the U.S. Rutabagas, taking over 90 days, are grown in northern areas where summer temperatures average 75 degrees F. or less.

Direct-sow about ½ inch deep in rows as close as 15 to 18 inches apart. Keep them well-watered and growing fast.

General soil and nutrient requirements are about the same as for beets, though these crops perhaps need slightly less nitrogen. Seedlings come quickly and easily, so the seedbed need not be extremely fine.

Varieties

Turnips: 'Early Purple-Top Milan', flattened 3 to 4-inch roots, 45 days;
—'Just Right', a white hybrid for greens or roots, 37 days;
—'Purple Top White Globe', the standard variety for roots, 58 days;
—'Seven Top', usually planted for harvesting greens in late fall or early spring, 45 days;

Turnip

—'Tokyo Cross Hybrid', makes 2-inch pure white roots but can grow larger without becoming pithy, All-America Selection, 35 days.
Rutabagas: 'American Purple Top', an old-timer, roots are purple above ground and light yellow below, yellow flesh, 88 days;
—'Macomber', very sweet, fine-grained, stores well, 92 days.

How to Use Turnips and Rutabagas

Turnip greens go well in salads when young and tender. They can also be steamed or boiled, and are especially good cooked with other greens such as mustard and chard.

The root should be peeled before using. Slice and serve it raw, or simply boil and serve with parsley and lemon butter. Baked, mashed, scalloped, or country-fried turnips are excellent with pork or game. Turnips can also be used in soufflés, soups, stews, and casseroles. Oriental cooks use them in stir-fry dishes and for making pickles.

Peel rutabagas before cooking, and, if cooking in water, add a teaspoon of sugar to improve their flavor. Rutabagas can be baked, French-fried, glazed, boiled, and mashed. They are good creamed with minced onion and seasoned with Worcestershire sauce, or as the prime ingredient in a vegetable soufflé.

Try seasoning rutabagas with nutmeg, fresh mint, dill, or basil. Tops can be cooked just like turnip greens.

SHALLOTS
See the Onion Family.

SOYBEANS
See Beans.

SPINACH
Spinach comes from Iran and adjacent areas. It spread to China by 647 A.D., Spain by 1100 A.D., and to America with the first colonists.

Spinach is a problem in many home gardens. The biggest obstacle is its tendency to bolt—to hurry into the flowering phase—which stops the production of usable foliage.

Bolting in spinach is controlled by day-length and highly influenced by temperatures and by variety selection. Long days hasten flowering, an effect increased by low temperatures during early growth and high temperatures in the later stage.

These circumstances make spring culture of a variety susceptible to

Spinach has a short life before going to seed.

bolting almost sure to fail; therefore, bolt-resistant (or long-standing) varieties must be used in spring. Some of the quick-bolting varieties that are otherwise good can be used in the fall, and in mild areas even in winter.

Make spring plantings in northern areas as early as possible, fall plantings about a month before the average date of first frost. In mild-winter areas plant any time from about October 1 to March 1.

In addition to bolting tendency, you may need to consider varietal resistance to two important diseases: downy mildew, or blue mold; and spinach blight, or yellows. Varieties also differ by having either savoyed (heavily crinkled) or smooth leaves. Savoy types are harder to clean but are dark green, thick, and usually preferred.

Varieties
Resistance to disease in the following varieties is indicated as follows:

Blight (B); Downy Mildew (DM); and Mosaic (M).

Long-standing: 'America', a savoy type, dark green, 50 days;
—'Long Standing Bloomsdale', a savoy, dark green, slow to bolt, 48 days;
—'Melody Hybrid', a medium savoy, All-America Selection, 42 days, (DM), M;
—'Winter Bloomsdale', smooth leaves, 45 days, (B).

Fall and winter only: 'Dixie Market', 40 days (DM);
—'Early Hybrid No. 8' (B, DM);
—'Hybrid No. 7', 42 days, (B, DM);
—'Virginia Savoy', 42 days.

SUMMER SPINACH
In the summer months when cool-season spinach fails the gardener, warm-season tropicals are available that are as rich in vitamins and as comparable in flavor to true spinach. They are:

Malabar spinach (*Basella alba*): When weather warms, this attractive, glossy-leaved vine grows rapidly to produce edible shoots in 70 days. Train it against a fence or wall. Young leaves and growing tips can be cut throughout the summer. Use cooked or fresh in salads. Seed sources are (5, 25, 31).

New Zealand Spinach (*Tetragonia expansa*): this low-growing, ground-cover-type plant spreads to 3 or 4 feet across. The young tender stems and leaves can be cut repeatedly through the summer. Seeds are really bundles of seeds, like beet seeds, and are slow to germinate. Start indoors in peat pots and set out after frost in spring.

How to use: Spinach may be served raw and crisp in salads, or alone with dressing and a garnish of crumbled bacon and diced hard-cooked egg. For the traditional spinach salad, wilt the leaves with a hot dressing of bacon fat, vinegar, mustard, chopped green onions, and honey, then add crumbled bacon.

The trick to preparing good spinach is quick cooking in as little water as possible — what clings to the leaves after washing is enough. Cover the pot and cook for 5 or 10 minutes until tender. Drain and season with butter, salt and pepper, and, if desired, a touch of lemon juice.

Serve poached eggs on a bed of spinach, or add it to omelets. Try it in a soufflé; or use it creamed as a crepe filling or for green pepper or tomato stuffing.

Prepare Malabar spinach like true spinach, and enjoy the hardy greens throughout the summer. They're good cooked or served raw in salads, and are popular in Oriental dishes.

SQUASHES, PUMPKINS, AND GOURDS
These are the members of four species of the gourd family, or *Cucurbitaceae* to botanists. (The entire family and its relationships are outlined in the chart on page 86.) All are native to the Americas. Most of our pumpkins and squashes originated in Mexico and Central America and were used all over North America by the Indians. Most of our winter squashes originated in or near the Andes in northern Argentina.

How to grow: Squashes and pumpkins, except for the bush kinds, are space users, not to be grown by mini-space gardeners lacking ways to use vertical space, such as training up a fence or trellis. Even a compost pile can serve as a place for the vine to ramble. They are sometimes also grown with corn, but in that case should be sparsely spaced.

On the ground, vining types need 10 feet or more between rows, but can be grown in less space by training or pruning. Long runners may be cut off after some fruit sets if a good supply of leaves remain to feed the fruit.

Like their relatives, squash and pumpkins are warm-season crops. Plant them when the soil is thoroughly warm, usually a week after the average last frost. Direct seeding is best, but if your season is too short to mature a direct-sown crop, use transplants from nurseries or start your own in individual pots 4 to 5 weeks before time to plant out safely. For the best crops, use hot caps or row covers to reduce transplant shock.

Leave 2 to 4 feet between plants, depending upon the vigor of the variety. Bush types do best in rows 5 to 6 feet apart, but can be grown as little as 16 to 24 inches apart in the row. We will provide specific directions on spacing for varieties differing from the norm.

Fertilizing and watering requirements are the same as for cucumbers and melons. Squash and pumpkins need generous amounts of organic matter in the soil, and 2 pounds of 5-10-10 fertilizer to 50 feet of row. Watering should be slow and deep. Leaves may wilt during midday, but should pick up again as the day cools.

Don't worry when the first blossoms fail to set fruit. Some female flowers will bloom before there are male flowers for pollen, and so will dry up or produce small fruits that abort and rot. This is natural behavior, not a disease. The same thing happens when a good load of fruit is set and the plant is using all of its resources to develop them. The aborting of young fruits will occur as a self-pruning process.

Harvesting: Pick summer squashes when they are young and tender. The seeds should be undeveloped and the rind soft. Pick continuously for a steady supply of young fruit. Zucchini and crookneck types are usually taken at 1½ to 2 inches in diameter, and bush scallops at 3 to 4 inches across.

Winter squashes are those that develop a hard shell when fully mature and can therefore be stored. They must be thoroughly mature to have good quality. When picked immature they are watery and poor in flavor. Flavor is usually better after some cold weather increases the sugar content. Learn to judge your varieties by color. Most green varieties get

Top: Crookneck
Middle: 'Golden Nugget'
Bottom: Zucchini

some brown or bronze, and 'Butternut' must lose all its green and turn a distinct tan.

Squash Varieties

From a long list of acceptable varieties, the following stand as good representatives of their classes:

Summer squash: Zucchini now come in yellow as well as shades of grey, green, and black. These four are All-America Selections (AAS):

—'Aristocrat', dark green, smooth, cylindrical, 50 days;

—'Chefini', early, dark green, solid, 48 days;

—'Gold Rush', deep gold color, good flavor, 52 days; and

—'Greyzini', grey, high-yielding, 50 days.

Also from the zucchini lists are 'Ambassador Hybrid', 48 days;

—'Burpee Golden', 54 days; and

—'Burpee Hybrid', 50 days.

Other summer squash include 'Blackini', early, bush, dark green, smooth, eaten when 6 to 8 inches long, 45 days;

—'Early Prolific Straightneck', big, bush, light yellow, cylindrical, 52 days;

—'Early White Bush Scallop', 60 days;

—'Golden Summer Crookneck', 53 days;

—'Patty Green Tint', a rich scallop type, 52 days;

—'St. Pat Scallop', early, vigorous producer of patty pan type (fluted rind) fruits, harvest at silver dollar size, 50 days;

—'Scallopini', dark zucchini color, scallop shape, productive, All-America Selection, 50 days;

—'Seneca Prolific', 51 days;

—'Sweet Mama', for small gardens, large fruits are dark grey-green, short vines are pinched off at 4 feet, cooks and stores well, AAS, 85 days.

Winter squash: 'Banana', long, pink and grey, 110 days;

—'Buttercup', sweet, strong flavor, 100 days;

—'Butternut', noted for mild flavor, 95 days;

—'Gold Nugget', bush, very early, 85 days;

—'Hubbard', large, good keeper, 110 days;

—'Acorn Table King', glossy dark green fruit, keeps well, AAS, 75 days;

—'Blue Hubbard', blue-grey, 15 pounds, AAS, 120 days;

—'Early Butternut', early hybrid, semi-bush, productive, AAS, 92 days;

—'Table Ace', acorn type, semi-bush, 78 days.

Winter squash will keep longer if they are placed on a ventilated shelf in a cool, dry spot like the garage or basement.

Zucchini, the most prolific of the summer squashes, has probably inspired more recipes than any single vegetable. The "zucchini glut" that strikes many gardens in midsummer can be handled in many delightful ways. Try sweet zucchini bread, pancakes, or cake. Split large ones and fill with ground beef, curried lamb, rice, seasoned crumbs, or any other stuffing, and bake, topping with herbs and grated cheese for the last few minutes in the oven.

Slice summer squash to cook in stir-fry dishes, quiches, and omelets. Crooknecks look festive halved, steamed in butter, and dusted with nutmeg. The pretty patty pan squash also takes well to stuffing. For barbecues, lay summer squash slices on foil, drizzle with olive oil, and add salt and pepper, garlic powder, and oregano. Grill about 20 minutes 5 inches above the coals.

With winter squash, the first step is cutting it down to serving size. Big varieties may require a heavy knife or handsaw for the initial cut. Then make sure to remove all seeds and stringy portions.

Winter squash can be steamed, baked, or broiled. To speed baking, put it on the pan cut-side down, cook till nearly tender, then turn right side up, add butter and seasonings, and bake another 15 to 20 minutes. Eat straight from the shell or scoop it out and dress it up with cream, brown sugar, and butter. Winter squash takes well to sweet spices and seasonings and many garnishes.

Pumpkin can be prepared by steaming or baking, as with winter squash, then seasoned with butter, salt and pepper, brown sugar or molasses. Try pumpkin bread, pie or cake, glazing either with corn syrup seasoned with lemon rind and ground ginger. For a great treat on a nippy fall day, bake pumpkin cookies filled with raisins.

Spaghetti squash: A unique winter squash, 'Vegetable Spaghetti' has string-like flesh that can be fluffed out of its shell after boiling and served as an excellent low-calorie substitute for spaghetti. It has medium-size oblong fruit and takes about 100 days to mature.

Pumpkins: 'Big Max', wins prizes for size, not eating, often over 100 pounds, 120 days;

—'Big Moon', 1 or 2 *huge* pumpkins per vine, each may weigh over 200 pounds, 120 days;

—'Cinderella', bush, medium-size jack-o-lantern, 95 days;

—'Connecticut Field' (or 'Big Tom'), large jack-o-lantern, 120 days;

—'Funny Face', early, semi-bush, 10 to 15 pounds, 97 days;

—'Green Striped Cushaw', unusual, pear-shaped, creamy white rind, pale yellow flesh, edible part is the neck, 90 days (4, 26);

—'Jack-O-Lantern', medium-size, 110 days;

—'Spirit Hybrid', medium-size pumpkins on vines more compact than most, symmetrical fruits perfect for Halloween carving, AAS, 100 days;

—'Small Sugar' or 'New England Pie', 100 days;

—'Triple Treat', uniform for carving, good flesh for eating, produces hull-less seeds for eating, 100 days.

Pumpkins for seeds form no hulls around seeds. In Mexico, pumpkin seeds are sold by street vendors, much as peanuts here. Two recommended varieties are 'Lady Godiva', with 8-inch fruits, 110 days; and 'Hungarian Mammoth', 115 days. Sources (5, 12, 19 and 27).

How to Use Squashes

Summer squash is good either cooked or raw. Use slices of zucchini or on a relish tray as dippers, or shred them and mix with sour cream dressing for a colorful slaw.

Spaghetti squash

French pumpkin

GOURDS

We have three reasons for these words on gourds: first, gardeners coming up with strange vegetable creations from their supposedly sincere vegetable patches and asking, "What is it?"; second, the increasing number of "edible" gourds finding their way into the vegetable seed catalogs; and third, letters from irate gardeners claiming we've misled them about cucumbers not crossing with melons.

This is why we asked a botanist into the picture to tell us where in the family of gourds the *edible* and *inedible* varieties belong, and to answer the question of crossing.

The chart on page 86 is the result. In some ways it's surprising: We'll never again have the same confidence in the common names "squash" and "pumpkin."

If you look down the "genus" column in the chart, you will come upon *Lagenaria* and its one species, *siceraria*. These are the larger gourds, often listed in seed catalogs as separate varieties such as bottle, dipper, and 'Hercules Club'. In many regions of the world, such gourds made up the total of domestic utensils before the invention of pottery, being fashioned into bottles, bowls, ladles, spoons, churns, and many other containers. They are still used to make musical instruments, pipes, and floats for fishing nets.

Seed growers have maintained the individual shapes and colorings of these ornamental gourds over the years by isolating the growing area for each. Different varieties will cross readily if grown together, and with time, present forms would change and new ones appear.

All gourds are fast growers if they have their quota of heat, especially warm nights. Delay planting until soil is warm. In short-season areas start seed indoors in pots 3 to 4 weeks before the average last frost. Set transplants (or thin to) 2 feet apart. Soil, water, and fertilizer needs are the same as for their relatives, the squashes, melons, and cucumbers.

Drying gourds: Most gourds can be dried. Pick them when the stems turn brownish. Punch the end close to the stem with a long needle to let air inside; then hang for several months in a well-ventilated place. The seeds will rattle when the gourds are fully dry. To make containers, cut them with a sharp saw and scrape out the insides; clean with a pot scrubber and cover inside and out with several coats of shellac.

Gourds peeking out from under their leaf cover.

Edible Gourds

Seed companies offer three edible gourds that we believe to be the same vegetable, a *Lagenaria siceraria* variety.

Catalog (4) calls it 'New Guinea Butter Vine' and describes it like this: "Will grow to enormous size— often 3 to 5 feet long and weighing 15 pounds. Similar to squash in growth, or can be trellised. Fruits should be eaten when small while the fuzz is still on them. Cook like squash or fry like eggplant."

Catalog (7) calls it 'Italian Climbing Squash', "an edible species of running gourd."

Catalog (19) calls it *Lagenaria longissima* and describes it as: "Italian vegetable used like summer squash if picked half-ripe. Has rich, full flavor. Delicious baked with fresh tomatoes, sprinkled with basil and olive oil. A Bostonian wrote, 'They are so good, I eat them for breakfast.'"

All the same vegetable? The sources we found are (4, 7, 11, 12, 19 and 21).

How to use: For the best eating, the edible gourds, or lagenarias, should be harvested young, while the fuzz is still on them. They have a rich, full flavor and can be cooked just about any way you would prepare summer squash or eggplant. Lagenarias are particularly popular in Italy; so try them Italian-style baked with fresh tomatoes and sprinkled with basil and olive oil.

Luffa

The luffa (*Luffa cylindrica*)—also known as the vegetable sponge, the dishrag gourd, and Chinese okra— grows fast to 10 to 15 feet, making cylindrical fruits 1 to 2 feet long.

Luffas are thought to have originated in tropical Asia. They reached China about 600 A.D., and are now cultivated throughout the tropics. Although tropical, the best luffas are grown in Japan.

To get the spongy, fibrous interior, the ripe gourds are immersed in a tank of running water until the outer wall disintegrates. They are then bleached and dried in the sun. Luffa is grown commercially for use as sponges and in the manufacture of many products—filters in marine and diesel engines, bath mats, table mats, sandals, and gloves.

In India the young, tender fruits are eaten like cucumbers or cooked as a vegetable. According to *The Gourd* magazine, in Hawaii and China the small pods are used to replace edible-podded peas in chop suey.

Look for luffa seed from sources (12, 19, 21, 25, 26, 31, 44, 47, 65, 68).

How to use: Luffas must grow to maturity for sponges, but for eating they should be picked when no more than 4 or 5 inches long.

Slice them raw into salads or butter-steam like summer squash. They adapt to most zucchini recipes, and combine especially well with tomatoes. Use the leaves in salads, or cook as greens. The flowers can be dipped in batter and deep-fried.

To make homegrown scrubbers, soak luffas in water for several days until the skin falls off. Then dry them in the sun.

Snake Gourd

Although the snake gourd, or serpent cucumber, grows up to 6 feet in length, it's best for eating when harvested at a fraction of that size. Snake gourds taste like cucumbers when eaten raw. When cooked they resemble zucchini in flavor and go well in almost any recipe for summer squash. They are a prized delicacy in Asia.

'Turk Turban'
This squash appears in both the decorative gourd section of the catalogs and in the winter squash varieties. It is widely available.

Vegetable Gourd
Closely related to both squash and pumpkins, this gourd is best grouped in the *pepo* species. A vigorously growing vine, it is attractive in foliage and fruits when trained on a trellis. Fruits are shaped like miniature pumpkins, 3 to 5 inches across, and weigh about ½ pound. When mature the gourd is striped a creamy white with dark green mottling. (11, 12, 19, 21).

How to use: Vegetable gourds taste like sweet winter squash. They can be stuffed like bell peppers with meat or rice, then baked; or boiled and mashed like winter squash.

Crossing in Gourds
Except for perhaps the mustard family, the gourd family has, among the vegetables, the greatest diversity in its edible forms, and certainly the widest variation in color and form of fruit. Cross two varieties of summer squash such as zucchini or yellow crookneck with white bush scallop, and the second generation will produce an unbelievable array of color, shape, texture, and size of fruit. If a large enough population is grown, there can be hundreds, of which no two are alike.

The pumpkin and squash seeds you buy are from varieties grown in areas free from the pollen of any other variety. However, nature has a way of sneaking in a cross or two. These will show up as occasional strange plants when commercial seed is planted in the garden. Seeds saved from these odd ones will produce many different forms the next year.

The possibilities for crossing can be summarized as follows:

Any two or more varieties *of the same species* will cross freely. For example, 'Hubbard' and 'Buttercup'

The Cultivated Members of the Gourd Family

FAMILY	GENUS	SPECIES	VARIETY OR COMMON NAME	
Cucurbitaceae	Cucurbita	pepo*		'Jack O'Lantern' pumpkin 'Connecticut Field' pumpkin 'Acorn' or 'Table Queen' squash 'Vegetable Spaghetti' 'Zucchini', 'Yellow Crookneck', and Bush Scallop summer squash Small, hard shell gourds Edible gourd Vegetable gourd
		moschata*	'Butternut' squash 'Kentucky Field' pumpkin 'Dickinson' pumpkin 'Golden Cushaw' pumpkin	
		maxima*		'Buttercup' squash 'Hubbard' squash 'Banana' squash 'Sweetmeat' squash 'Marblehead' squash 'Turks Turban' squash 'Big Max' pumpkin 'King of Mammoths' pumpkin
		ficifolia	Malabar gourd	
		mixta*	'Green Striped Cushaw' pumpkin 'White Cushaw' pumpkin	
	Lagenaria	siceraria	Bottle or White Flowered gourd Cucuzzi	
	Luffa	cylindrica	Dishrag gourd	
	Momordica	balsamina	Balsam apple	
		charantia	Balsam pear	
	Sechium	edule	Chayote	
	Benincasa	hispida	Chinese Preserving melon or White Gourd of India	
	Cucumis	melo		Netted muskmelon or Cantaloupe ('Hales Best', 'Harper Hybrid', etc.) Honeydew & Casaba muskmelons (Honeydew, Crenshaw, etc.) Oriental Pickling melon Dudaim melon Snake or Serpent melon Mango melon
		sativus	Cucumber, all varieties	
		anguria	West Indian Gherkin	
	Citrullus	vulgaris	Watermelon, all varieties Citron	
	Trichosanthes	anguina	Snake or Serpent gourd	

*See text about crossing between these species

squash will cross, as will 'Harper Hybrid', 'Hales Best', and 'Crenshaw' melons.

Crossing between species does occur in the genus *Cucurbita. Pepo* will cross with *moschata* and *maxima* will cross with *moschata. Pepo* and *maxima* will not cross. An additional cross, *pepo* with *mixta*, will occur also. All of this means, for example, that 'Acorn' will cross with 'Butternut', but 'Acorn' will not cross with 'Hubbard'.

Other crosses between species, such as muskmelon with cucumber, do not occur. Nor do any crosses occur between one genus and another.

Salsify
See the Root Crops.

Sugar peas, or snow peas
See Peas.

Sweet potatoes
This member of the morning glory family *(Ipomoea)* was taken from Central and South America to Spain by Columbus. After the conquest of Mexico, the Spaniards took sweet potatoes to the Philippines, and the Chinese adapted them there. Records show that they were being cultivated in Virginia in 1648.

No vegetable commonly grown in the United States will withstand more summer heat, and very few require as much heat, as the sweet potato. This tropical plant does not thrive in cool weather: A light frost will kill the leaves and soil temperatures below 50 degrees F. will damage the tubers.

Commercial crops are feasible where mean daily temperatures (average of day and night) are above 70 degrees F. for at least three months. Louisiana, North Carolina, and Georgia all produce significant commercial crops. In the West, only the hottest-summer areas of Arizona and California support commercial culture.

Although the names "sweet potato" and "yam" are often used interchangeably they are different plants, and differ in both growth habit and culture. Yams belong to the genus *Dioscorea* and are rarely grown outside of the tropics. They are not as nutritious as sweet potatoes, being little but starch. Shredded and cooked, they become mucilaginous, or gluey.

Fertilizing sweet potatoes is tricky. Given too much nitrogen, they develop more vines than roots. However, they are not a poor-soil crop, and a low-nitrogen fertilizer such as 5-10-10 worked into the soil at the rate of 4 pounds to 100 feet of row will improve yield. Prepare the soil two weeks before planting. In some areas potash will increase yields and make roots shorter and chunkier; low potash likely means long, stringy potatoes.

How to plant: Sweet potatoes are started from slips. Supermarket tubers may sprout and produce the slips, but like Irish potatoes they are usually treated to prevent sprouting. Disease-free slips from a garden center or mail-order seed company are best, and you'll be sure of your variety.

Plant promptly after the soil is warm, since length of growing season is often the limiting factor. Where the season is long enough, the experts recommend planting 2 weeks after the last frost to insure thoroughly warm soil.

If drainage is poor or plants are overwatered, roots may be elongated and less blocky. To insure good drainage, sweet potatoes normally are planted in 6 to 12-inch ridges, located 3 to 4½ feet apart. (However, southern gardeners with fast-draining sandy soil report that ridge planting for them is no great advantage.)

Space the plants 9 to 12 inches

Sweet potatoes

apart. Rich soil may allow closer spacing. If spacing is increased much beyond 24 inches, more oversized roots will result, but at the expense of both total yield and quality.

Being deep-rooted plants often grown in moisture-poor sandy soils, sweet potatoes are usually able to find enough water to survive; but for harvest, they should never have to struggle for water. As a general guide, they will use about 18 inches of water per season.

After your first sweet potato crop, save some of the healthiest tubers to produce your own slips. Kansas State University recommends:

"Cut or split the crown and underground stems of each plant and examine for dark strands or for general darkening of internal tissue, a symptom of stem rot infection. Do not use roots from stems whose internal tissue shows any discoloration, even though the sweet potatoes appear okay. Also examine for surface cracks and black 'eyes', which indicate nematode infestation."

Three potatoes should produce about 24 slips, enough for a 25-foot row and about 60 pounds at harvest. Plant the chosen tubers close together in sand, vermiculite, or perlite hot beds 5 to 10 weeks before your outdoor planting date. When sprouts reach 9 to 12 inches, cut them off 2 inches above the soil, and set the slips into a good, preferably sterile rooting soil. (We used 4-inch plastic containers filled with a peat/perlite soil mix.) Keep the soil warm (80 to 90 degrees F.) for fast rooting. Sufficient roots should take 10 to 14 days.

Harvesting: Harvest sweet potatoes when slightly immature, if the size is adequate. Otherwise wait until the vines begin to yellow. Try to avoid bruising them when digging as this invites decay. If the leaves are killed by frost, harvest immediately.

Sweet potatoes improve during storage because part of their starch content turns to sugar. For storage, they need to be cured. Let the roots lie exposed for 2 to 3 hours to dry thoroughly, then move them to a humid (85 per cent relative humidity) and warm (85 degrees F.) storage area. After two weeks lower the temperature to 55 degrees F., and they will keep between 8 and 24 weeks.

Beginners' mistakes: Planting too late to take advantage of the full growing season is the most common error. But don't plant too early either. If soil is not at least 50 degrees F., plants will languish, if not perish.

Limited space: If you lack the gar-

From top: Yellow Pear, 'Golden Boy' hybrid, pear-shaped 'San Marzano', and 'Snowball'.

Varieties: 'Jersey Orange', 'Nugget', and 'Nemagold' are the popular dry-fleshed varieties. 'Centennial', 'Porto Rico', and 'Gold Rush' rate high in the list of moist-fleshed kinds.

How to use: Sweet potatoes in this country means two basic types of vegetable: the pale yellow, slightly sweet, fairly dry northern version; and the dark orange, moist, distinctly sweet southern type. The southern "sweets" are those commonly called yams, and we'll call them that here, though they are not true yams.

Most cookbooks use the two types interchangeably, but common sense dictates going easier on sweeteners with yams and heavier with butter or cream on the drier sweet potatoes. Dieters should be aware that yams are much higher in calories.

Both vegetables are delicious baked in the skin and served with butter. Glaze them in brown sugar or maple syrup, or with orange juice (and a little rind) or crushed pineapple. Try baking boiled, sliced sweet potatoes with sliced apples, garnished with raisins and chopped nuts.

Sweet potatoes or yams complement pork chops, lamb, turkey, or ham, and also find their way into a number of baked treats, from sweet potato pie to biscuits to cake.

Tomatoes

None of the vegetable gifts from the New World to the Old World took as long to be appreciated as the tomato. Used for centuries in Mexico and South America, it is recorded as being cultivated in France, Spain, and Italy in 1544; but a century later it was being grown in England only as a curiosity.

The first seeds to reach Europe were of the yellow variety. They became the *pomi d'oro* ("apples of gold") of Italy, and a few years later, the *pommes d'amours* ("apples of love") of France.

In pioneer America only a few would venture to eat the fruit, which was thought poisonous by many. New Englanders in Salem in 1802 wouldn't even taste them. But by 1835 they were recognized by the editor of the *Maine Farmer:* "a useful article of diet and should be found on every-man's table."

Tomatoes are warm-season plants and should be planted at least one week after the average last frost. Temperature is a most important factor, and tomatoes are particularly sensitive to night-time temperatures. In early spring when day temperatures are warm but nights fall below 55 degrees F., many varieties will not set fruit. In summer you can expect blossom drop when days are above 90 degrees, or nights above 76 degrees F.

Check the growing-season charts on pages 28-33 to find the best times for you to plant, and choose varieties that do well in your climate.

Soil for tomatoes should be well drained and have a good supply of nutrients, especially phosphorus. To prepare soil, use plenty of organic matter, adding in 3 to 4 pounds of 5-10-10 fertilizer per 100 square feet. Water it in and allow two weeks before planting.

Most gardeners start tomatoes with transplants, which are usually available at garden centers at the earliest possible planting time. To start your own transplants, sow seed ½ inch deep in peat pots or other plantable containers 5 to 7 weeks before the outdoor planting date. See page 46 on sowing transplants. The last ten days before planting out, gradually expose the seedling to more sunlight and outdoor temperatures.

In the garden, tomato is a hard-to-germinate seed, but you'll find a valuable tip on page 44.

Transplants should be stocky, not leggy, and should have 4 to 6 true leaves, young and succulent. Avoid plants already in bloom or with fruit, especially if growing in very small containers.

Set transplants deep, the first leaf just above soil level. Plant leggy plants with the root ball horizontal. Roots will form along the buried stem and make subsequent growth better. If cold or wind are threats, use hot caps or other protection.

The first fertilizer application will take care of the plant until it sets fruit. Feed then and once a month while the fruits are developing, and stop when they near mature size.

Tomatoes require uniform moisture after fruit has set: alternate wet and dry spells can bring on stunting and blossom-end rot. In the early stages, if careful not to overdo it, you can stretch watering intervals to put the plant under a little stress—it's a good way to bring on tomato production.

Tomatoes in containers: Tomatoes are ideal container plants for everyone, and gardeners with unfavorable soils especially find that growing in containers with a disease-free planter mix is worth the extra care in watering and feeding.

Grow dwarf plants in 8-inch pots. Several varieties can be grown in hanging baskets. Giant-size plants will thrive even in 2-gallon containers

den space for sweet potatoes, but would still like to grow some, try them in a box at least 12 inches deep and 15 inches wide. Use a light, porous soil mix and place a 4-foot stake in the center to support the vine. Or grow them as a lush, vining houseplant indoors in a bowl or jar.

of soil if you compensate for the limited root space with extra water and fertilizer. Use any container you like, but be sure to provide drainage.

Any tomato that can be grown in a vegetable garden can also be grown in a container.

Training: No vegetable responds better to training than the tomato, and few plants are trained as easily. They can be grown on upright stakes, trellises, and in wire cages, or on horizontal trellises or ladder-like frames set a foot above ground level. All will keep fruits from contact with the soil and reduce damage from slugs, cracking, sunscald, and decay. In wet fall climates the yield of usable tomatoes can nearly be doubled by so protecting the fruit.

Every gardener we have worked with has given us a "best way" to train tomatoes. There are probably many "best ways." Probably the favorite method for large-growing, main-season varieties is the circular cage made from concrete reinforcing wire. A 5½-foot length will make a cage 18 inches in diameter, and the 6-inch mesh allows for easy picking. The yield of quality fruit is higher in cages than on stakes.

The low-growing, bushy varieties are difficult to stake but may be held up by horizontal frames of different kinds. (They may be allowed to sprawl, but should then be protected by mulch, organic or plastic.)

Tall-growing varieties may be grown on 6-foot stakes set a foot deep into the soil.

Pruning: The market gardener aiming for the early market will sacrifice quantity and prune to a single stem, removing all suckers. This shortens the growing season and ripens fruit uniformly.

The home gardener can modify this system and get more fruit over a longer period by allowing one sucker to grow from near the base to form a two-stemmed plant, and later removing the rest of the suckers on both stems.

One disadvantage of heavy pruning is a lack of foliage to shade the fruit and protect against sunscald. Pruned plants are also more prone to develop blossom-end rot.

A way to get early fruit production and later sun protection is to take out all suckers up to about 18 inches, then let the plant bush out with the branches tied to the stake.

Protection: Early planting calls for protection against temperatures. If you use the wire cage, a cover of polyethylene film in the early stages of growth will raise temperature.

The covering of a row of 4 or 5 plants with the film, as illustrated, may be used with a row tent only 2 feet high, or, later, with a tent as high as the mature plant.

Tomato Problems

All vine and no fruit: It's often been said that the cause of "all vine and no fruit" is too much nitrogen and water in early growth stages. Plants too liberally supplied with both certainly will make a lot of vine, but the real explanation is that a plant remains all vine simply because no fruit set. If blossoms drop, tomatoes will continue growing in the vegetative stage.

Blossom drop: For a tomato grower this can create great anxiety. The blossoms are out, but the big question is, will they drop or set fruit? It takes about 50 hours to find out—the minimum time required for the pollen to germinate and the tube to grow down the pistil to the ovary. At night temperatures below 55 degrees F., germination and tube growth are so slow that blossoms drop off before they can be fertilized. As a rule, most early-maturing varieties set fruit at lower temperatures than the main-season kinds.

Transplanting

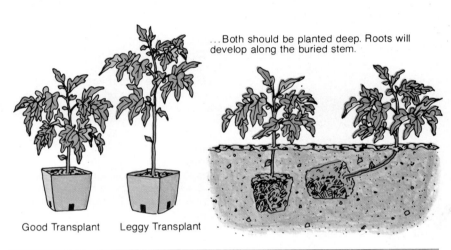

...Both should be planted deep. Roots will develop along the buried stem.

Good Transplant   Leggy Transplant

Staking and Training

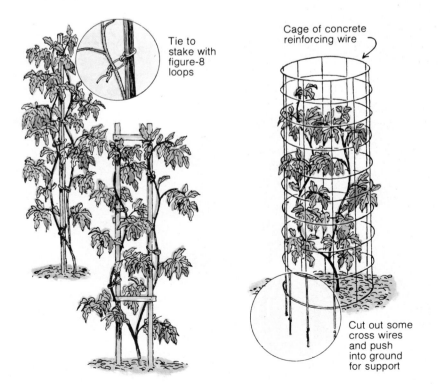

Tie to stake with figure-8 loops

Cage of concrete reinforcing wire

Cut out some cross wires and push into ground for support

'Patio' tomato

'Pixie' tomato

Blossom-end rot: Symptoms of this disease appear as a leathery scar or rot on the blossom-end of fruits. It can occur at any stage of development and is usually caused by sudden changes in soil moisture, most serious when fast-growing plants are hit by a hot, dry spell. Lack of calcium is another cause.

Mulching with black plastic or organic material, which reduces fluctuations in soil moisture and temperature, and avoiding planting in poorly drained soil, will help prevent blossom-end rot. Also, staked and heavily pruned tomatoes seem more susceptible to the disease than unpruned plants.

Curled leaves: Wilt, during a hot spell at midday, is normal. Plants in containers show top-growth wilt and drooping when they need water. Once watered, their speed of recovery is amazing.

Some kinds of leaf curl are normal. It's more pronounced in some varieties, and you can expect it during hot, dry spells and during and after a long wet period. Heavy pruning also seems to encourage leaf curl.

Failure to set fruit: We have already explained the role of temperature in fruit set. It can also be hampered by rain or prolonged humid conditions. Growers in cool, humid situations have found that fruit set can be increased by shaking the plant, or vibrating it with a battery-powered toothbrush, to release pollen for pollination. When plants are trained on stakes, hitting the top of the stakes will have the same effect.

Poor fruit color: In hot summer areas, high temperatures can prevent the normal development of fruit color. The plant's red pigment does not form in temperatures above 86 degrees F. Uneven coloring is common if fruits mature in high temperatures.

Both high temperatures and high light intensities will stop the color from forming in fruit exposed to the direct sun, and fruits may sunscald. Where high temperatures are the rule, choose varieties with a dense foliage cover.

Beginners' Mistakes
Failure to fit the variety to the climate is one way to a disappointing crop.

Failure to choose disease-resistant varieties may be a cause of failure.

Planting too early in the season is a common mistake.

Trying to grow tomatoes in a shady location gives poor results: they require at least 6 hours of direct sunlight.

Tomato Varieties
The rule of thumb in choosing tomato varieties to fit your garden is this: The shorter your growing season, the more you should limit your choices to the "early" and "early midseason" varieties. Finding them should be no problem, as you can see by looking at the variety chart on pages 92-93.

In checking the varieties listed in the following chart, note especially those with resistance to soil-borne pests. Their resistance is indicated by the initials "V" for *verticillium*; "F" for *fusarium*; and "N" for *nematodes*. It may be that your soil is not infested with any of these pests and you can successfully grow any variety. But if you've had any trouble with tomatoes in the past, favor the resistant varieties.

The number of days shown on the chart refers to the time from setting out transplants to the first fruits. It is an average figure and intended only as a general guide.

The growth habit of the variety is indicated by the words "determinate" or "indeterminate." The determinate are the bush kinds, generally growing to 3 feet or less. The indeterminate are tall-growing and are trained with stakes, trellis or wire cage.

How to Use Tomatoes
To many tomato lovers, the supreme tomato is the one plucked from the vine on a midsummer afternoon and eaten right in the garden. For other gardeners, growing that tomato is only a beginning: they're intrigued by the limitless possibilities of tomatoes in the hands of the good cook.

Ripe tomatoes should not be kept

Determinate/Indeterminate

Determinate. These are commonly called bush or self-topping tomatoes. The terminal buds set fruit and stop the growth of the main stems.

Indeterminate. Terminal buds do not set fruit, just more leaves and stem. The vine will grow indefinitely if not killed by frost.

Here are two more ways to train tomatoes. The clear plastic offers wind protection to young plants.

in the refrigerator for any length of time. You'll get top flavor if you store them in a cool place, as close to 60 degrees as possible. To peel, cover them with boiling water for 10 seconds, immerse in ice water until cool, then remove the skin with a serrated knife. Try not to peel or cut tomatoes until just ready to use.

Raw tomatoes can be sliced and served on an antipasto tray; marinated with sliced cucumbers in a vinaigrette dressing; combined with zucchini, hard-cooked eggs, and chili peppers; or added to any tossed salad. Or stuff them with such fillings as tiny shrimp, potato salad, cold salmon, seasoned cottage cheese, or mashed deviled egg.

To insure the best color, flavor, texture, and food value, cook tomatoes until just tender in a small amount of water; tomatoes create their own liquid as they simmer, and there's no point in diluting the nutrients. Cooked tomatoes are used in omelets, stews, casseroles, meat sauces, salad dressings, and even breads. You can brew your own tomato juice, create a secret spaghetti sauce, or make fresh cream of tomato soup. Or enjoy gazpacho, a spicy cold soup of tomatoes, cucumbers, and bell peppers, topped with sour cream.

Tomatoes are also excellent cooked as a single vegetable dish. Try them stewed, scalloped, stuffed and baked (spinach makes a tasty filling), or

sliced and glazed in wine and brown sugar. For barbecues, skewer tomato slices and grill them quickly over the coals.

With the danger of approaching frost, many gardeners find themselves with an abundance of green tomatoes—which must be used quickly or not at all. Don't hang them upside down in the garage to ripen or rot, but channel them straight to the kitchen. You'll soon realize you've discovered a totally new vegetable.

Green tomatoes are delicious sliced, dipped in batter and breadcrumbs, and fried in hot oil. They make tangy relishes such as piccalilli, which can be enjoyed all winter, or can be turned into mincemeat for use in cookies, cakes, and pies. Add them to orange marmalade for tartness or make simple green tomato jam. To many cooks, the best dish of all is green tomato pie, often compared to rhubarb or apple in flavor. Serve this distinctive dessert warm and topped with vanilla ice cream.

Putting up canned tomatoes or tomato preserves is one of the best ways to handle a sudden tomato glut. Canning is a safe, straightforward process if you find good canning instructions and follow them. Much has been written in the press about the risk of botulism from home-canned tomatoes, particularly from varieties low in acid content. The USDA has collected considerable data in an ef-

fort to ascertain the possible dangers.

According to its report, "The seriousness of botulism should never be underestimated. However, we believe that the risk to home canners of tomatoes is very small. The home canner should not be overly concerned about hazards associated with the selection of specific tomato varieties. It is far more important to select tomatoes which are not overripe, to follow the recommendations of reliable canning guides explicitly, and destroy (without tasting) any home-canned product which appears abnormal in any way."

For insurance that the tomatoes you can are in the low pH range (or high in acidity), add ¼ teaspoon citric acid or 1 tablespoon lemon juice per pint of tomato product.

TURNIPS AND RUTABAGAS
See the Root Crops.

YARDLONG BEAN OR ASPARAGUS BEAN
See Beans.

WATERMELON
See Melons.

ZUCCHINI
See Squash.

TOMATO VARIETIES	Hybrid	Days to maturity	Growth habit	Fruit size	Disease resistance	Comment
WIDELY ADAPTED						
'Better Boy'	×	72	Ind	Large	VFN	Main crop, 12 to 16-ounce uniform fruit. Heavy cropper, rugged.
'Big Set'	×	75	Det	Large	VFN	Sets well in both high and low temperatures, vigorous, resists cracking, catfacing and blossom-end rot.
'Bonus'	×	75	Det	Med.-Lg.	VFN	Smooth, firm, catface-resistant fruits. Sets well in high temperature.
'Burpee's Big Boy'	×	78	Ind	Ex.-Lg.		Smooth, firm, thick-walled, productive.
'Burpee's Big Early'	×	62	Ind	Large		Early and large fruit, vigorous vine.
'Burpee's Big Girl'	×	78	Ind	Large	VF	Same as 'Big Boy' but with VF resistance.
'Burpee's VF'	×	72	Ind	Med-Lg.	VF	Widely adapted and recommended, vigorous vine and meaty fruit.
'Early Cascade'	×	61	Ind	Med.	VF	Early, tall, and productive.
'Early Girl'	×	62	Ind	Med.-Sm.	V	Productive, vigorous vines, fruits firm and meaty.
'Fantastic'	×	70	Ind	Med.-Lg.		Productive, widely adapted and recommended.
'Florimerica'	××	72	Det	Large	VF	All-America Selection, tolerance or resistance to 17 diseases or disorders.
'Heinz 1350'		75	Det	Med.-Lg.	VF	Strong, compact vine, uniform 6-ounce fruit.
'Marglobe'		75	Det	Medium	F	Long-time favorite; firm, smooth, thick-walled 6-ounce fruit.
'Monte Carlo'	×	75	Ind	Large	VFN	Tall, strong vine, productive over long season.
'Rutger's'	×	80	Semi-det	Med.-Lg.	F	Descendent of original 'Rutgers' and similar. to 'Ramapo', widely available.
'Springset'	×	65	Det	Medium	VF	Vigorous, open vine; susceptible to sunscald; good crack resistance.
'Spring Giant'	×	65	Det	Large	VFN	First All-America tomato, 1967; high yielding, concentrated harvest season.
'Super Sioux'		70	Semi-det	Medium		Noted for high-temperature fruit setting ability; best in wire cage.
'Terrific'	×	70	Ind	Large	VFN	Strong-grower, produces over long season; good cracking resistance.
'Vineripe'	×	80	Ind	Large	VFN	Heavy yields, vigorous grower.
'Wonder Boy'	×	80	Ind	Med.-Lg.	VFN	Strong vine, medium foliage cover, heavy yield.
BEEFSTEAK TYPES WITH WIDE DISTRIBUTION						
'Beefmaster'	×	80	Ind	Large	VFN	Beefsteak type with triple disease resistance.
'Beefsteak'		90	Ind	Large		Ribbed, irregular and rough fruit; vigorous vine with coarse foliage.
'Oxheart'		86	Ind	Large		Heart-shaped fruits to 2 pounds; pinkish, firm, meaty and solid.
'Pink Ponderosa'		90	Ind	Large		Old-timer, meaty and firm; grow in cage.
BEEFSTEAK TYPES WITH LIMITED DISTRIBUTION						
'Brimmer'		85	Ind	Large		Smoother than regular beefsteak; meaty with few seeds and no core (32).
'Bragger'	×	87	Ind	Large		Strong-grower, red and meaty fruit; prune to one stem and stake or cage (6, 9, 21, 67).
'Burpee's Supersteak'	×	80	Ind	Large	VFN	Consistent texture, shape and size; good for slicing (5).
'Pink Gourmet'	×	70	Det	Large	F	Large, firm, pinkish fruits; first fruits rough, later ones more smooth (18, 51).
SMALL TOMATOES						
'Gardener's Delight'		85	Det	¾"		Prolific and crack resistant; one of the sweetest (17).
'Pixie'	×	52	Det	1¾"		Early fruit on 14 to 18-inch plants; right in 8-inch pot or hanging basket (5, 27).
'Small Fry'	×	60	Det	1"	VFN	Cherry-type in clusters on 30 to 40-inch vine; All-America Selection 1970; widely available.
'Sugar Lump'		70	Det	1"		Cherry-size fruit on 30-inch vine for hanging basket or short trellis; unusually sweet (21).
'Sweet 100'	×	62	Ind	1"		Very large, multiple-branched clusters of very sweet fruit; prune to one stem and stake or cage (7, 9, 10, 17).
'Tiny Tim'		55	Det	¾"		Small, scarlet fruit on 15-inch vine; two in an 8-inch pot or hanging basket; widely available.
'Toy Boy'	×	68	Det	1½"	VF	Plant 3 or 4 in a single 10-inch pot; indoors or out with plenty of light (4, 7, 9, 10, 19, 67).
'Tumblin' Tom'	×	72	Det	1½"		Early and heavy yield from 1 to 2-foot plants; good for hanging basket or windowbox.
PASTE TOMATOES						
'Chico III'		75	Det	Sm.-Med.	F	Compact and slightly open-growing, good for juice, sets fruit at high temperatures.
'Roma VF'		75	Det	Small	VF	Strong grower with dense foliage cover.
'Royal Chico'		75	Det	Small	VF	Growth habit improved compared to 'Roma'.
'San Marzano'		80	Ind	Small		Larger, rectangular, straight-sided pear; drier than most.

TOMATO VARIETIES	Hybrid	Days to maturity	Growth habit	Fruit size	Disease resistance	Comment
UNIQUE VARIETIES						
'Stakeless'		78	Det	Med.-Lg.	F	Very dense foliage more like relative potato, thick-stemmed 18-inch plants; support when loaded with fruit (6, 27, 28).
'Climbing Triple-L Crop'		90	Ind	Large		Will keep growing to 10 or 15 feet, beefsteak-type fruit.
SPECIAL VARIETIES FOR THE SOUTH						
'Atkinson'		77	Ind	Medium	FN	From Auburn U., firm and meaty fruit, good protection from foliage.
'Floradel'		77	Ind	Med.-Lg.	F	From Florida AES, for commercial, home garden and greenhouse; sets well at low temperatures.
'Floralou'	×	74	Ind	Medium	F	Smaller fruit than 'Floradel' but will out-yield it.
'Homestead 24'		82	Det	Medium	F	Sets under variety of conditions including high temperatures.
'Manalucie'		87	Ind	Medium	F	Firm, meaty fruit; vine vigorous and upright; a southern favorite.
'Manapal'		75	Ind	Medium	F	Bred for humid growing climate by Florida AES; a productive southern favorite.
'Marion'		79	Ind	Medium	F	One of the best home-garden stake varieties, widely recommended.
'Nematex'		75	Det	Medium	FN	From Texas AES, bush-type vine gives good cover, used wherever nematodes a problem.
'Porter'		65	Ind	Med.-Sm.		An old reliable that consistently produces over wide range of soil and climatic conditons.
'Porter Improved'		65	Ind	Medium		Same vigor and reliability with improved fruit color and size.
'Traveler'		78	Ind	Medium	F	From Arkansas AES, pink fruit, popular commercial tomato.
'Tropi-Red'		77	Det	Large	VF	From Florida AES, compact bush-type vine;
'Walter'		75	Det	Medium	F	Vigorous and reliable, must be harvested ripe; stake or let sprawl.
SPECIAL VARIETIES FOR THE NORTH						
'Jet Star'	×	72	Ind	Med.-Lg.	VF	Important variety in the second early season, sets well, good production.
'Moreton Hybrid'	×	70	Ind	Large		Old-time favorite; red color, good texture and flavor.
'New Yorker'		67	Det	Medium	V	Sturdy plants reliably produce heavy crops.
'Ramapo'	×	85	Ind	Med.-Lg.	VF	From Rutgers U., sets well under adverse conditions, strong and vigorous.
'Supersonic'	×	79	Ind	Large	VF	Reliable main cropper in northern areas.
SPECIAL VARIETIES FOR THE WEST						
'Ace 55'		76	Med.-det	Large	VF	Adds disease resistance but reduces fruit smoothness compared to standard 'Ace'.
'Royal Ace'		80	Det	Large	VF	Averages larger fruits than standard 'Ace'.
'Cal-Ace'		80	Semi-det	Large	VF	Sets heavily, fruit smooth.
'Hybrid Ace'	×	80	Det	Large	VFN	Adds resistance to nematodes.
'Ace-Hy'	×	76	Det	Large	VFN	Smooth fruit; roots resist nematodes.
'Early Pak 7'		82	Lg.-det	Med.-Lg.		From Ferry-Morse Seed Co., compact vine; cage, stake or trellis.
'Early Pak 707'		82	Lg.-det	Med. Lg.	VF	More uniform, reliable fruit plus disease resistance.
'Fireball'	×	65	Det	Medium		Compact and dwarf plant; concentrated, short harvest season.
'Hotset'		72	Ind	Med.-Lg.		Sets fruit when daytime temperatures are as high as 90 degrees F., good in Southwest.
'Improved Porter' ('Porter's Pride')		65	Ind	Medium		Improved descendent of one of the best for Southwest.
'Jetfire'	×	60	Det	Medium	VF	Larger and better fruit with better foliage cover compared to 'Springset'.
'Manitoba'		60	Det	Medium		Similar to 'Bush Beefsteak' but smoother, vines compact, medium foliage cover.
'Nematex'		75	Det	Medium	FN	Heavy yield of quality fruit, used in Texas soils with nematodes.
'Pakmore'		75	Det	Large	VF	Heavy production of large fruit, grow in cage, 'Pearson'-type.
'Pearson Improved'		90	Det	Large	VF	Somewhat open vines with good disease resistance.
'Pearson A-1 Improved'		90	Det	Med.-Lg.		Smooth but slightly smaller fruits, good foliage cover, favored in California and Arizona.
'Porter'		65	Ind	Medium		Consistently produces over wide range of climatic and soil conditions, good heat resistance.
'Saladette'		65	Det	Small	F	From Texas AES; makes heavy crop of tasty, firm and crack-resistant fruits under adverse conditions.
'Starfire'		56	Det	Med.-Lg.		An improved 'Fireball' type; compact, bushy vine.
'VFN 8'		75	Det	Med.-Lg.	VFN	Holds up well after fall rains, from UC California.
'Willamette'		67	Det	Medium		Always tops in Oregon trials, some crack resistance.
'Young'		72	Semi-det	Med.-Lg.	F	Productive; resists cracking, blossom-end rot, catfacing, and puffing.

VEGETABLES OF SPECIAL INTEREST

Some vegetables are common in one region but not in another; some are rare and unusual nationwide. This chapter describes the cultural requirements for each one and may help you expand your vegetable garden possibilities.

In most ways, these vegetables reflect regional differences. Okra is common in southern gardens, but is "unusual" to many northerners. The situation is similar for artichoke, jicama, chayote, and watercress: common in some areas, rare in others. (But even more remarkable is the *wide* adaptation of so many of our common vegetables.)

These regional lines are becoming less distinct. Thanks to air transportation, jicama is frequently available in New York produce markets, far from its native Mexico. And many determined gardeners are proving that the range of other rarer plants is not as restricted as we might think.

Some of these vegetables are as easy to grow as weeds. In fact, you might previously have considered them only as weeds (see *Dandelions,* page 97). The culture of others is involved enough to demand real interest.

The 59 catalogs listed on pages 104-105 are our source material. We read each and cross-checked among them all. If the vegetable in question was infrequently mentioned, we noted by number the nurseries offering it. Many, however, are widely available and so noted.

The National Academy of Sciences has published a book titled, *Underexploited Tropical Plants with Promising Economic Value.* Two plants described in it have received considerable attention:

Winged bean (Psophocarpus tetragonolubos) is a short-day plant and has been successfully grown in southern Florida and parts of the Gulf Coast. It has been glorified by Noel Veitner of the NAS as follows:

"The winged bean is a veritable backyard supermarket. All but the stringiest parts of the plant are edible. Pods and seeds are steamed and boiled. Leaves are used as spinach. Growing shoots and stems are cooked as asparagus. Tubers are eaten as potatoes. The flowers are also edible. Besides all this food-producing ability, the plant improves the soil when used as a cover crop and is perennial."

If you want to try this one, it's available from sources (68, 70).

Grain amaranth is a vegetable that is certainly not new, but is in the process of rediscovery and getting much attention. Gurney Seed &

An inside view of an artichoke reveals the beautiful symmetry, but it gives no idea of its delicious flavor. This vegetable is now becoming more widely distributed with improved shipping methods.

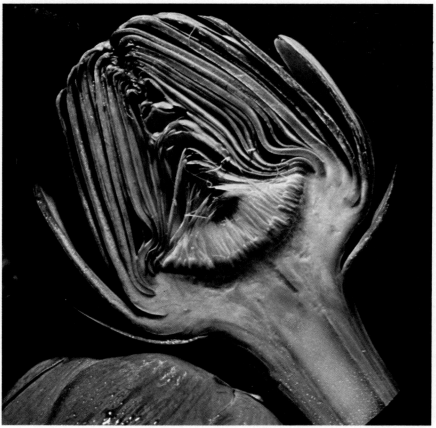

Here are some of the special interest vegetables. **1.** Horseradish **2.** Chayote **3.** Watercress **4.** Jicama **5.** Tomatillo **6.** Jerusalem Artichoke **7.** Peanuts **8.** Rhubarb **9.** Belgian Endive

Nursery Co. has this to say about it:

"A grain crop of the Aztecs, the plant produces round, white seeds (sesame seed size). It makes a nutritious flour and is high in protein and other nutrients. Yield is 1 to 2 ounces per plant or 1,000 pounds per acre. It is a rare and rightfully prized grain."

If this one interests you, the sources are (12, 44).

In the next few pages, we describe other unusual plants that you may find valuable both in your garden and in the kitchen.

ARTICHOKES

Centuries before Christ the Romans were paying top prices for this thistle relative and preserving it for year-round use. It disappeared with the Roman Empire and did not come to light again until a thousand years later in southern Italy. Catherine de Medici found artichokes in Florentine gardens and introduced them to France as gourmet delicacies. And it was French and Spanish explorers who brought them to this country.

Commercial artichoke production is primarily restricted to the cool, humid, moderate climate of the California coast from San Francisco south to Santa Barbara. However, a variety known as 'Creole' grows in southern Louisiana; and artichokes also have been produced by determined gardeners in northern states. It is important in cold-winter areas to mulch the crown for protection, but not so heavily as to smother it.

Generally, artichokes do not produce true from seed—you may or may not get good artichokes. Most are therefore planted from root divisions. A full-size clump will produce 3 or 4 plants, or divisions may be found at your local nursery. Seeds are available, however; Thompson & Morgan (17) offer artichoke 'Grande Beurre', about which they say: "At last a variety of artichoke that produces large fleshy heads of a consistent size."

Start seeds in spring about the time of the last frost, sowing about ½ inch deep in the garden; or start them indoors 4 to 6 weeks before the average last frost.

Plant root divisions 6 to 8 inches deep, 4 to 6 feet apart in the row. Leave 7 or so feet between rows—the plants can get big. Plant to a side of the garden where they won't be in the way of the more frequent planting and cultivating of annual vegetables, and allow for some afternoon shade if you're in a hot-summer area.

In areas of year-round production, feed in the fall with a high nitrogen fertilizer. In cold-winter areas, feed in spring when new growth starts with about a pound of 10-10-10 per plant.

Slugs and snails are common pests of artichokes. Try hand-picking, barriers, traps, and baits. The artichoke plume moth lays eggs in the bud. Where these are most troublesome, a regular insecticide program may be needed.

Artichoke sources are (5, 7, 10, 12, 14, 15, 21, 22, 25, 26, 31, 36, 40, 65, 70).

How to use: Artichoke buds are boiled or steamed until tender (about 45 minutes) and brought to the table either hot or chilled. The leaves are pulled one at a time and, usually, dipped into a sauce; then the tender inner flesh is eaten by pulling the leaf through the teeth. The rest of the leaf is discarded.

When the leaves are all removed, scrape away the fuzzy center with a spoon and enjoy the heart, considered a delicacy. Hearts may also be marinated, or served in a lemon, oil, and herb sauce.

Cardoon

CARDOON

This handsome plant, with its silvery fernlike foliage, is a favorite with Italian and French cooks. Because of its ease in growing and preparation, it deserves more recognition in American cuisine.

Closely related to the artichoke, the cardoon has a flavor somewhat between celery and zucchini. Physically it resembles the artichoke, with large, ornamental 3 to 4-inch thistles, deeply cut leaves, a crown that multiplies by sending out side-branches, and a heavy flower head complete with the thistle's purple bristles. But while the artichoke is raised for its

fleshy flower bud, cardoon is grown for young leafstalks.

Plant cardoon in the garden where it can be appreciated close-up. A cool-season plant, it requires 120 to 150 days from seed to harvest and needs rich, moist soil. Keep plants well fed and watered for vigorous growth—if they struggle to survive, leafstalks will become pithy and the plant will put its energy into flowering.

Harvest by cutting off the blanched stems just below the crown. Then trim off the outside leaves. Remaining will be a blanched heart some 18 to 24 inches thick.

Sources are (7, 14, 19, 25, 36, 46, 50, 65, 67, 71).

How to use: Cut cardoon stalks into sections and parboil in salted water and lemon juice (to prevent darkening) until tender. Serve in a cold salad with a vinaigrette dressing; or as a hot vegetable seasoned with butter, cheese, or a light cream sauce. Italian cooks prefer it dipped into a light egg batter and deep-fried until just crisp.

CHAYOTE

Chayote, also known as "vegetable pear" and "mirliton," is a member of the gourd family, but certainly doesn't look like a gourd. In mild-winter areas it grows as a perennial. Frost will kill back the tops but the vine renews itself in spring. It is fast-growing and best handled on a trellis or fence. Flowers appear in late summer and fruit is harvested about a month later, continuing until frost.

Plant chayote in the spring after all frost danger. The whole fruit is used as the seed. Place it on a slant with the wide end down, stem end slightly exposed. The vines are vigorous and grow quite large—one plant can produce 3 dozen fruits, more than enough for most home gardens.

In cold climates, mulch the roots heavily with compost or similar material for winter protection. Pull the mulch aside in spring at sprouting time. The growing plant will need plenty of water and fertilizer, but go easy with nitrogen: too much will produce excessive growth.

Store chayote fruits in a cool place. They will keep for 2 or 3 months for later eating or for seeding in the spring. If the plant sends out shoots in storage, which is likely, cut them back to 2 inches when you plant.

Chayote is not started from seed, but you can plant fruits from the market.

How to use: Young chayote can be cooked without peeling. Large, fully mature fruit will have a tough skin. Cut into slices, right through the flat inner seed. It has a nut-like flavor after cooking.

Chayote can be used in more ways than zucchini. It may be diced and steamed until tender, baked and stuffed, cooked and marinated for use in cold salads, or pickled and candied. A favorite food of Mexican cooks, it takes seasonings well and complements most every dish.

CRESS

Four kinds of cress are cultivated, and more are found growing wild.

Early winter cress, or Belle Isle Cress: This is a hardy biennial which survives severe winters. Sow it in late summer and use it in summer before the seed stalks develop, and in winter. More bitter than watercress, it should be cooked with another vegetable, like spinach, to cut its strong flavor.

Garden Cress, or Pepper Grass, Curly Cress: This is a fast-growing annual with the look of parsley. It germinates in a few days if the seeds are exposed to light, and can be eaten in 2 weeks as sprouts, or later in all stages up to maturity. It is a cool-season, short-day plant best grown in early spring or for fall harvest.

Garden cress seeds require light to germinate. We know a kindergarten teacher who uses this trait in her teaching. Seeds in two 6-inch pots are planted ½ inch deep. These will not germinate. However, in another container a child scratches his or her initial ½ inch deep and simply lays the seeds inside it. The initial grows.

Upland Cress: This dwarf plant with slender stalks and oval, notched leaves resembles watercress in shape and flavor. It makes dense growth 5 to 6 inches high and 10 to 12 inches wide. Sow seeds in rows 12 to 14 inches apart. Thin to 4 or 8 inches between plants and use the thinnings. Young, tender tips will be ready in about 7 weeks. Upland cress requires moist soil and is tolerant of fall frosts.

Watercress: This is the best known and most widely used commercial cress. The commercial product is a perennial grown in pure, gently running water. If you have a small stream or spring suitable for watercress, you'll find detailed information on growing it there by sending 10¢ for *Commercial Growing of Watercress* (Farmers Bulletin No. 2233) to the Superintendent of Documents, Washington, D.C. 20402.

In the home garden adaptation of its natural growing conditions, you can grow watercress from either seeds or cuttings. Make cuttings from market watercress, stick them in sand or planter mix in a pot, and place the pot in a tub of water. Or sow seed in small containers and transplant when seedlings are 2 or 3 inches tall. Plant out in a planter box, cold frame, or a trench in the soil, wherever plants can be given a continuous supply of water. Cress sources are (1, 3, 5, 7, 13, 14, 20, 21, 25, 32, 36, 38, 40, 41, 44, 45, 46).

How to use: In Britain, cress is a very popular green. The British Ministry of Agriculture says it is one of the very few foodstuffs which are almost balanced diets by themselves.

Cress makes a spicy, fresh addition to a tossed green salad, and an effective garnish with hot or cold dishes. The French have popularized watercress soup, a thick soup with a cream-of-potato base. Italian cooks add cress to minestrone. The Chinese have long used it in won ton.

DANDELION

This is the same basic plant that grows as a weed in your lawn, but the cultivated kinds will grow much larger, with thicker leaves and better flavor.

The dandelion is a cool-weather vegetable, so plant in early spring or late for fall harvest. If you've ever collected wild dandelions, you know the best are found in moist, fertile soil in the cool of spring.

The dandelion is easily grown from seed. It is perennial and if allowed will come up each year.

Pick greens when the leaves are still young and tender and before the plant flowers. With advanced maturity they become too fibrous and tough. To slow this, tops can be blanched by tying up the leaves or by covering the plant to exclude any light.

Sources are (1, 5, 14, 15, 19, 21, 27, 36, 39, 41, 42, 46, 47, 50, 65, 71).

How to use: Called spring-tonic greens, dandelions contain more iron and Vitamin A than any other garden vegetable or fruit.

The leaves, especially of the cultivated varieties, have a tangy taste that goes well in salads with thin slices of sweet onion and tomato and chopped basil. Or they can be cooked and eaten like spinach. Add greens to a thick lentil soup along with bits of bacon and chopped onion. The roots are roasted and ground and used like coffee.

Dandelion wine is a time-honored tradition in many gardens. Said to be among the better of the home-

Chayote

Dandelion

Watercress

made wines, the bouquet is similar to champagne, and overall, the effect memorable.

Chicory

ENDIVE, CHICORY, ESCAROLE

When buying these lacy, slightly bitter greens at the market, you can call them "chicory" or "endive" as you please. They are sold under both names. But when you want to plant the frilly one, buy endive *(Cichorium endiva)*.

For the blanched, tubular French or Belgium endive, buy seeds of witloof chicory *(Cichorium itybus)*.

Endive is grown in the same way as lettuce but is at its best if grown for fall or winter harvest. The finely cut, furled kinds are represented by the varieties 'Green Curled', 'Pancalier', and 'Ruffec'.

The less frilled, broad-leaved type of endive, often called 'escarole,' is catalogued as 'Full Heart Batavian' and 'Florida Deep Hearted'. Both types are best for salads when blanched by drawing the outer leaves together and tying with a string. In 2 or 3 weeks blanching should be complete.

While endive and escarole are grown much like lettuce, witloof chicory (French or Belgian endive) is a bit more involved. In cold-winter areas the usual method for this famous winter salad crop is to grow it from early summer to fall. Then, after cutting the tops, set the roots in moist soil in a warm cellar and cover with a 6-inch layer of moist sand. With this forcing, new leaves grow in the sand and produce a tight, blanched head.

In mild-winter areas, the same end can be achieved in the garden. Sow seed in early summer—not too early, because a plant that goes to seed is

useless for forcing. Thin to 4 to 6 inches apart. When plants are fully grown and the first frost hits, cut off the tops 2 inches above the crown to prevent injury to the crown buds. Cover the roots with 6 to 8 inches of soil. In late December start harvesting by removing soil to expose the white shoots in the first few feet of the row. After cutting the shoots, scrape the soil back over the root in a mound to force the second crop. Drive a stake to mark how much row you have harvested. Harvest can continue through winter and early spring.

Varieties: 'Cicoria Catalogna' and 'Radichetta', both used for early greens, leaves toothed and curled, flower shoots edible with a faint asparagus flavor, 65 days;

—'Cicoria San Pasquale', used for greens, light green leaves broader than 'Magdeburg', 70 days.

—'Large-Rooted Magdeburg', about 15 inches tall, upright dandelion-like leaves, root 12 to 14 inches long, young tender leaves can be harvested for greens at 65 days, roots mature at 120 days.

Sources: (11, 25, 40, 41, 42, 46, 63, 65).

FLORENCE FENNEL, OR ANISE

The vegetable that supermarkets commonly call "anise" or "sweet anise" is known to gardeners as Florence fennel. It's also known as *fenouil* to the French and *finocchio* to the Italians, and we call it "sweet fennel" or just plain "fennel." Fennel was very popular among the Romans, who it is said served virtually no meats or vinegar sauces without it.

The plant grows about 2 feet high with broad leafstalks that overlap each other at the base, forming a bulbous enlargement that is firm, sweet, and white inside. There is a common, or weedy, fennel that grows 4 to 6 feet high. Its seeds are useful for flavoring, but it lacks the broad stalks at the base.

Fennel requires cool weather. Plant it either as early in spring as the soil can be worked, or in summer for a fall crop. Mild-climate areas plant it fall for a spring harvest.

Sow seeds about ¼ inch deep in a sunny spot. Thin to 18 inches apart. When plants are half-grown, hill

up soil around the bases to blanch.

Sweet fennel will wait in the garden a considerable time until you're ready to harvest.

How to use: Fennel has a delightful licorice-like flavor. Use it much like celery. Cut off the green stalks and tough outer leaves. Slice thinly, lace with olive oil, and season. Chill before serving. Excellent with fish, fennel is often added to court-bouillon for seafood or to basting sauce for broiled fish.

GARDEN HUCKLEBERRY

If you happen to read that "this is the cultivated form of the poisonous black nightshade", change it to read "edible form."

This relative of the potato, eggplant, and tomato grows about 2¼ feet high. The ½ to ¾-inch berries are borne in clusters and are shiny black when ripe. They are not edible until black and soft. Culture is the same as for eggplant and tomato.

Left: Green curled endive.
Below: Florence fennel.

Sources: (9, 10, 12, 16, 20, 26, 36, 39, 65).

How to use: The garden huckleberry sometimes has a bitter taste which can be removed by parboiling for 10 minutes in water containing a pinch of baking soda. When combined with lemons, apples, or grapes, it makes excellent jellies and preserves.

Recipe for huckleberry pie: Wash and stem 2½ quarts of berries, cover with water and let come to a boil. Add ½ teaspoon soda, boil 1 minute and drain. Add 1 cup cold water, cook until soft, then mash the berries and add 1½ cups sugar and the juice of ½ lemon, and boil about 15 minutes. Remove from the stove and cool. When ready to put the filling into the crust, add 1 tablespoon of tapioca. Dot with butter, add the top crust, and bake.

HORSERADISH

Horseradish grows naturally throughout much of eastern Europe from the Caspian through Russia and Poland to Finland. Planted in Colonial American gardens, it escaped to flourish as a wild plant.

Garden huckleberry

Horseradish rarely produces seed and is generally grown from root cuttings. Ask for them from your local nurserymen, or check our catalog sources.

Set the root cutting small end down and large end 2 to 3 inches below the soil surface. Keep plants about 12 inches apart. Roots set out in spring will be harvest-size in fall. When leaves are about a foot high, pull back the soil above the cuttings and remove all but 1 or 2 of the crown sprouts. At the same time, rub off (using gloves) the small roots from the side of the cutting. Don't disturb the branch roots at the base. Re-cover the root with soil. This operation can be repeated in about a month if you want top-quality horseradish.

Most growth occurs during late summer and early fall, so it's best to delay harvest until October or November.

Sources: (4, 5, 9, 11, 12, 15, 16, 19, 21, 26, 50, 51).

How to use: Beloved for its tangy flavor, horseradish plays a role in cuisines throughout the world. Peel and grate the root directly into white wine vinegar or distilled vinegar. (Do not use cider vinegar, which discolors horseradish within a short time.) The vinegar may be full strength or slightly diluted, as you prefer. Bottle as soon as possible after grating, and refrigerate at all times to preserve the pungent flavor. The mixture will keep for a few weeks. Horseradish may also be dried, ground to a powder, and bottled. So prepared, it will not be as high quality, but will keep much longer than when grated fresh.

Mix horseradish with whipped or sour cream to accompany sauerbraten, roast beef, pork, or tongue; or add it to seafood sauces for a special tang.

Jerusalem artichoke may be prepared in all the same ways as potatoes, but with an important difference: they should never be overcooked. Overcooking tends to toughen the vegetable.

They are delicious raw. Add slices to salads or use as a last-minute garnish in clear soups. Lightly boil the tubers, with or without their skins, in salted water until tender (usually 15 to 20 minutes); add them to salads with oil-and-vinegar dressing, or quarter and sauté slowly in butter until just tender.

JICAMA, OR YAM BEAN

This is an unusually interesting vegetable, new to American cooks but grown in its native Mexico for centuries. It is now increasingly being offered in U.S. supermarkets.

The plant, a vine, is grown principally for its large tuberous root, which is vaguely turnip-shaped and usually four-lobed. The skin is a brownish gray and the flesh crisp and white. The flavor resembles water chestnut but is sweeter. In fact, one seed company (19) actually lists jicama as water chestnut, adding, "The immature pods are edible but the leaves, ripe pods, and seeds are toxic and narcotic and should not be eaten by man or beast."

Jicama makes an attractive ornamental, worth a place in the flower garden. Its flowers are profuse, white to lavender in color, and resemble sweet peas.

Jicama is a tropical plant and requires at least 9 months of warm growing season to mature its large roots. If rich, light, friable soil and 4 months of warmth are available, roots will be small, but still quite tasty.

Note the planting instructions on page 48. Soaking seeds in water for 24 hours will hasten germination. Seeds may be sown as suggested for climbing string beans, or near a trellis for support.

Transplants should be set into the garden as soon as weather warms. When growth reaches about 3 feet, pinch the tips to promote horizontal branching. Tubers form as days begin to grow shorter, and should be harvested before first frost. Do not allow the plants to go to seed or the tubers will be small. Flowers, which appear in late summer, should be pinched out for the maximum root production.

Sources: (12, 15, 17, 25, 52, 65, 68, 70).

How to use: Jicama is delicious served as an appetizer: Peel and cut

HUSK TOMATO, OR GROUND CHERRY, STRAWBERRY TOMATO

Husk tomato, a bushy plant about 1½ feet tall, is grown in the same way as tomatoes, and the fruit is about the same size as the cherry tomato.

Fruits are produced inside a paper-like husk. (The ornamental Chinese paper plant is a close relative.) When ripe, the husks turn brown and fruits drop from the plant. Fruits left in the husks will keep for several weeks.

Sources: (4, 9, 10, 12, 16, 17, 19, 20, 25, 26, 39, 44, 47).

How to use: The fruit is sweeter than the small-fruited tomatoes, and is used in pies and jams or may be dried in sugar and used like raisins. Hawaii's Poha jam is husk tomato jam.

JERUSALEM ARTICHOKE, OR SUN CHOKE

This easily grown vegetable is native to eastern North America. It bears no relation to the artichoke, except vaguely in flavor, nor does it have any connection to the Holy Land. In 1616, explorers of the New World discovered it being eaten by the Indians and took it back to Europe under the Italian name for the sunflower, *Girasole* (literally, "turning to the sun"), and the French name, *Artichauts du Canada.* Today the French call them *topinambour,* and the English, *Jerusalem* (likely a corruption of the Italian *girasole*).

A hardy perennial, this plant is a species of sunflower, growing 6 to 10 feet high on a single stalk, and topped by 3-inch yellow sunflowers. The stalks are often used as a windbreak for tender crops or stripped of leaves and dried to serve as stakes for pole beans the following season. But the plant is grown primarily for its

Husk tomatoes (top and above) resemble Chinese lantern plant.

round, knobby root or tuber, which can be used like a potato.

Plant the whole tuber or cut pieces with 2 or 3 eyes. Dig a furrow 4 inches deep, drop in a tuber or piece every foot, and cover. Rows should be 3 feet apart. When the first foliage appears, start to hoe, more toward the plants than away, to hill up soil around them as with potatoes.

Tubers should be ready for harvest in about 100 to 105 days. Dig them as you use them. Leave in the ground for storage over the winter. We have found the plant a very exuberant grower, and even the smallest pieces of tubers left in the ground produce plants the following spring.

Sources: (4, 5, 7, 10, 12, 15, 16, 19, 21, 44, 71).

How to use: Unlike potatoes, Jerusalem artichokes are starch-free; their carbohydrates do not convert to sugar in the body, and they can be eaten by the diet-conscious and the diabetic without concern. Use the tubers as soon as they are dug—they keep only in soil or sand. Store as potatoes rather than refrigerate.

into strips or thick slices, sprinkle with lime juice and salt, and arrange on a tray. Add it to fruit salads or to raw vegetable salads. Or slice it thinly and prepare with pan-fried potatoes.

NASTURTIUM

If you have grown nasturtiums only for their contribution to the beauty of the garden, you are missing the important bonus they offer the good cook.

Everything you need to know about growing nasturtiums, according to many seed catalogs, goes like this: Very easy to grow, blooms profusely in ordinary, well-drained soil; even thrives in dry, sandy, or gravelly areas. Too-rich soil may cause plants to produce more leaves than flowers.

Varieties: Which nasturtium variety should you choose? Any one will meet kitchen requirements. If you intend to plant in the vegetable garden, training the long vines on chicken wire would take the least space. However, for boxes and pots there are many low-growing dwarf varieties. The foot-high 'Double Jewels' offers a choice of color. And if you like the simplicity of the old-fashioned kinds, there's a dwarf *single* nasturtium available. Sources for many varieties are (5, 7, 12, 16, 17, 19, 23, 26, 28, 32, 61, 65).

How to use: The lively, subtle, peppery taste of nasturtium leaves is reminiscent of watercress. Chop and add them to all kinds of salads. Flowers can serve as carriers of cheese mixtures or as salads for the hors d'oeuvres tray. The plump, green, unripened seedpods are often

Nasturtium salad

pickled in vinegar and substituted for capers.

OKRA

Okra, sometimes called "gumbo" (though the name gumbo is more properly applied to soups containing okra), achieved its popularity in the French cookery of Louisiana and was probably introduced to that region by French colonists in the early 1700's.

There are two types of okra: the tall growers, to 4 or 7 feet; and the so-called dwarfs, to about 3 feet. Okra is an attractive plant, being an edible form of hibiscus and having hollyhock-like flowers. A half-dozen plants will provide more than enough fruit over a long season for the cupful you'll need now and then for various dishes, and also leave enough for pickling.

Okra is a warm-season vegetable. Planting dates and fertilizing and watering schedules for corn apply to it also. Plant 8 to 10 inches in rows 3 to 4 feet apart. In short-season areas, start seeds in small pots about 5 weeks before you would plant corn or beans, and set out when the soil is thoroughly warm.

Harvesting: To keep the plant producing, no pod should be allowed to ripen on the stalk. Young pods are more tender and more nutritious. The pods develop rapidly and plants should be picked over at least every second day. Handle pods with care; do not break or bruise them or they are apt to become slimy or pasty during cooking.

Varieties: Among okra varieties, 'Clemson Spineless' grows to 4 or 5 feet and is a heavy yielder. 'Perkins Spineless' is a dwarf, 2½ to 3 feet high. 'Red Okra' produces rich red pods that turn green when cooked. If left to seed, the plant becomes a thick 5 to 6-foot bush. Pods picked at maturity can be used like typical okra or dried and worked into interesting and unusual floral arrangements. For various varieties, see sources (1, 5, 15, 21, 28, 32, 33, 34, 36, 38, 39, 40, 41, 42, 61, 67).

How to use: Whether boiled, baked or fried, okra should be cooked rapidly to preserve flavor and prevent the development of a slimy consistency. Do not cook it in iron, copper, or brass utensils. While harmless, the chemical reaction will discolor the pods.

The young, tender pods of okra are very popular in Creole cooking, and are excellent in soups and stews. They possess a natural thickening agent, and perform this function in gumbos. Okra combines well with other vegetables, especially tomatoes.

Okra can be steamed or boiled, served with a variety of sauces, or pickled (above).

PEANUTS

This tropical from South America was taken by the Portuguese from Brazil to West Africa. Spanish galleons carried peanuts from Southern America to the Philippines, from where they spread to China, Japan, and India. They found a favorable climate in North America early in our history. Thomas Jefferson wrote of their culture in Virginia in 1781.

There are two types of peanut: Virginia, with 2 seeds per pod; and Spanish, 2 to 6 seeds per pod. Most plants of the Virginia type are spreading, and the Spanish are bunching. However, bunching varieties of Virginia are available and better adapted to short-season areas than the spreading types.

A long, warm growing season of 110 to 120 days is required for most peanuts. If summers are cool, better forget them as a crop regardless of the season length.

The strange growth habit of peanuts inspires gardeners to experiment in growing them even when they must be in containers and given special protection. The plant resembles a yellow-flowering sweet pea bush. After the flowers wither, a stalk-like structure known as a "peg" grows from the base of the flowers and turns downward to penetrate the soil. When the peg pushes to a depth of 1 to 2½ inches, it turns to the horizontal position and the pod begins to form.

The soil should be light and sandy. Peanuts require a generous supply of calcium in the top 3 or 4 inches of soil where the pods develop. Foliage is often dusted with gypsum (calcium sulfate) at the time of flowering, at the rate of 2½ pounds per 100 feet of row.

See the chart on page 48 for specific planting directions.

When plants are about 12 inches tall, mound soil around the bases and cover with a mulch. Make sure plants get a regular supply of water up to 2 weeks before harvest. Excess water at harvest time may break dormancy and cause the mature peanuts to sprout.

Harvesting: When the plants turn yellow at the end of the season, lift each bush carefully with a garden fork, shake free of soil, and hang the plants, with the peanuts hanging by the pegs, in a warm airy place for a few days. Let the plants cure for 2 to 3 weeks before stripping the peanuts from them.

Peanuts are widely available.

How to use: Peanuts are so well known with so many uses, that it would be futile to attempt a listing here. But we frequently are asked about roasting. To roast peanuts with no chance of scorching, place them unshelled in a colander or wire basket. Preheat the oven to 500 degrees F. Place the peanuts in the oven and turn it off. When the peanuts are cooled to the touch, they're ready to eat.

RHUBARB

This hardy perennial is grown for its leafstalks. The leaves contain poisonous quantities of oxalic acid.

Rhubarb varieties adapted to the northern United States and Canada require two months of temperatures around freezing to break their rest period; and, for quality and yield, also need a long, cool spring. Plant in early spring as soon as the soil can be worked.

Start rhubarb from root divisions (rooted crowns) with 1 to 3 buds, or eyes. Rhubarb can be grown from seed but the results are variable and not the quality of selected crowns. Root divisions offered by your local garden store are picked for quality and climate adaptability.

Plant in a trench 12 to 18 inches deep and filled with a good soil mix to within 2 or 3 inches of the top. Set crowns about 2 inches below the soil surface. Do not allow them to dry out before planting.

Rhubarb needs space to grow—3 feet between plants. But 3 or 4 plants will supply all the average family can use. Locate the plants out of the way of regular gardening operations.

Rhubarb will be around for 4 to 6 years. The first year after planting allow all stalks to grow. The second year harvest only for 1 or 2 weeks. After that, you can have fresh rhubarb for 8 weeks or more.

Varieties: The popular varieties in cold-winter areas are 'MacDonald', 'Ruby', and 'Valentine Red'. In mild-winter areas, use 'Giant Cherry', which needs less cold. Rhubarb is widely available.

How to use: If young and tender, there is no need to peel stalks before cooking, just wash them and cut into 1-inch chunks. Older stalks, however, may need to be peeled and "de-strung" like celery.

Always cook rhubarb. Steam it in a double boiler or stew it like applesauce, adding sugar to taste near the end of cooking time. Many cooks spoil rhubarb by sweetening too much, which kills its natural tartness. Rhubarb is often combined with strawberries in salads or fruit compotes. Try it in a molded gelatin salad with cream cheese, chopped celery, and nuts. Alone or with berries, it makes good jelly and jam.

Rhubarb is best known in the American original, rhubarb pie—which comes in many forms: double crust, deep dish, meringue, or even a custard. Rhubarb is also popular in cakes and puddings; and spiked with port or brandy, it becomes a delicious topping for ice cream.

Rhubarb finds a place among flowers.

SHUNGIKU

Shungiku is often referred to as "chop suey greens," a more general name that also includes other plants of similar use. Another name is "garland chrysanthemum," which is somewhat more appropriate because the plant is an edible kind of chrysanthemum. The leaves are similar to those of chrysanthemums and it has bright yellow, daisy-like flowers in the fall.

Shungiku is grown as a perennial. Plant seed in early spring in rows 1 to 1½ feet apart. Or plant in solid beds as with carrots or lettuce. Once plants reach 5 inches tall, you can begin to harvest, either by pulling the whole plant to thin at the same time, or by just removing some leaves.

Sources: (11, 17, 19, 21, 25, 44, 45).

How to use: The luxuriant young growth made in the cooler parts of the year are best. The flavor is strong but not sharp or bitter. It can be used fresh in salads or as an addition to stir-fried vegetables.

SUNFLOWERS

Stand next to this garden annual when it is full-grown and you'll feel as if you've stepped into a land of giants. Growing as tall as 12 feet, with flower heads as large as 18 inches across, sunflowers stand like sentries, topped with huge sunbursts of color. Plant them at the back of the flower border

Sunflowers are a popular ornamental in many areas.

Tomatillo is similar to husk tomato.

or in the vegetable garden, and each will supply a bounty of seeds for good eating.

Some gardeners grow sunflowers as windbreaks, or plant the seeds two weeks before pole beans to let the stalks support the climbing vines.

Sow the seeds when soil is warm. (See the planting chart on page 48.) If you have a problem with birds snatching the seeds, cover the seedheads with cheesecloth and tie it securely at the back of the flowers.

Varieties: There are many sunflower varieties available, ranging from a few inches in height to over 12 feet. Some form a single large head and others many smaller heads. A new development is the early-maturing medium-size plants (5 to 6 feet) that form full-size heads (8 to 10 inches). One example is 'Sunbird Hybrid' (5). Other varieties are available from many sources.

How to use: A county agent in Washington State gave us the following recipe for roasting sunflower seeds. Hang the sunflower heads in a dry location with good air circulation. Tie a cloth bag around each head to catch any seeds that might drop during drying. When seeds are dry, mix thoroughly 2 cups of unwashed seeds, ½ teaspoonful Worcestershire sauce, 1½ tablespoons melted butter, and 1 teaspoon salt. Place in a shallow baking pan and roast for one hour at 250 degrees F. To ensure even browning, shake the pan several times during the process. Place seeds in a plastic bag and store in the refrigerator.

Squash and pumpkin seeds can be roasted the same way.

TOMATILLO

The tomatillo is a perennial, often grown as an annual, reaching 3 to 4 feet high. The leaves are long, oval, and deeply notched. The fruit is smooth, sticky, either green or purplish in color, 1 to 2 inches in diameter, and entirely enclosed in a thin husk.

A close relative of the husk tomato, the tomatillo has a tart flavor—to some, similar to green apples. Tomatillos and green tomatoes are never interchangeable in recipes. The flesh is of a different texture, seedy but solid, without the juicy cavities of the tomato.

Grow tomatillos the same ways as tomatoes. Seed sown in peat pots will germinate in about 5 days and be ready to transplant in 2 to 3 weeks.

Harvest tomatillos according to the intended use. For highest quality fruit, harvest fresh when the husks change color from green to tan; otherwise, they lose their tartness and become soft. Left on the vine, they become yellow and mild. They can be stored for months. Some gardeners store them on the vine; others spread out the picked fruit still in their husks, in a cool place with good air circulation. (Jammed into airtight plastic bags, the berries spoil rapidly.)

Sources: (47, 52, 65, 68, 73).

How to use: Tomatillos are used in many Mexican recipes. Raw or cooked, they give sauces a rich, distinctive flavor. Try them fresh in salads, tacos, and sandwiches. Tomatillos are probably best known for the contribution they make to Mexican *salsa verde*, or green taco sauce. Add chilis for hotness.

Seed Sources

All of the seed-grown vegetable varieties in this book are available from the companies on this list. To help you find the scarcer varieties, those with just one or a few sources, we have added numbers in parentheses telling you where to write for seeds or catalogs. For example, 'Little Leaguer Cabbage' (5) in the book refers you to (5) on the list, the Burpee Co.

You may notice that some numbers are missing from our list. This is because when companies go out of business or otherwise drop from the list, their numbers are also dropped. This way numbers are the same in the 1974 edition and this new edition, and remain consistent throughout all of the Ortho gardening publications.

All-America Selections

In your gardening pursuits, you may often have run across the phrase, "All-America Selection." Catalogs frequently cite these honored plants, and we've noted many in this book. Just how a plant wins the title and what it means are of importance to the gardener.

The All-America Selections organization includes a council of judges and some 50 test gardens located in different climates throughout the U.S. and Canada. Its purpose is to evaluate the new vegetable (and flower) varieties introduced each year. The entries, known to the judges by number only, are grown in trial rows side-by-side with the most similar varieties already in commerce. Judges vote primarily on home garden merit, paying special attention to climatic adaptation and vigor.

From the 75 to 100 entries each year, usually only 3 or 4 earn enough points to be given the All-America medals. The gardener thus can be assured that an All-America Selection is superior in its class and adapted to as wide a climate range as any.

Mail-order catalogs generally list the new All-America Selections before they appear on seed racks in nurseries and garden centers. However, some winners of past years are often not given their due; for example, 'Tender-Pod' beans, bronze medal, 1941; 'Top Crop' beans, gold medal, 1950; 'Emerald Cross' cabbage, silver medal, 1963; 'Batavian Full Heart' endive, silver medal, 1934; 'Salad Bowl' lettuce, gold medal, 1958; and 'Spring Giant' tomato, bronze medal, 1967.

Using Catalogs

Variety names may vary among catalogs even though the vegetables described are indistinguishable from one another. In our experience, the "fingerling" or container carrots known as 'Little Finger', 'Baby Finger Nantese', 'Lady Finger', 'Sucram Baby', 'Baby Nantese', and 'Tiny Sweet', are all the same carrot. A variety you are unable to find may be available somewhere under a different name.

Many mail-order companies have special interests: herbs, Oriental vegetables, melons, peppers, and many others. Some of these are noted in the list.

We have researched 59 catalogs in the making of this book. Many of them are valuable gardening books in their own right. Tucked in amidst their vegetable entries are cultural directions, useful hints, practical ideas, and even recipes, all based on years of gardening experience. With only a few catalogs, you can find much enjoyment, improve your gardening skills, and keep up with the world of vegetables.

---

(1) Allen, Sterling & Lothrop
191 U.S. Rt.1
Falmouth, ME 04105
Straightforward listing of varieties and prices. No descriptions.

(3) Meyer Seed Co.
600 South Carolina Street
Baltimore, MD 21231
Vegetables, 21 pages. Features All-America Selections. Special listing of outstanding selections.

(4) Burgess Seed & Plant Co.
Box 82
Galesburg, MI 49053
68 pages, 8½ x 11. Vegetables, 26 pages. Special attention to varieties for northern states. Many unusual items.

(5) W. Atlee Burpee Co.
Warminster, PA 18974;
Clinton, IA 52732; or
6350 Rutland Avenue, Box 748
Riverside, CA 92502
166 pages, 6 x 9. When seed catalogs are mentioned, most people think "Burpee."

(6) D.V. Burrell Seed Growers Co.
Box 150
Rocky Ford, CO 81067
96 pages, 8½ x 4¼. Seed growers. Special emphasis on melons, peppers, tomatoes and varieties for California and the Southwest.

(7) Comstock, Ferre & Co.
263 Main Street
Wethersfield, CT 06109
40 pages, 8½ x 11. Vegetables, 11 pages. An informative guide to variety selection. Over 50 herb varieties. Founded 1820.

(8) Jackson & Perkins
1 Rose Lane

Medford, OR 97501
The well-known name in roses adds the Jackson & Perkins Seedbook to their catalog list. 12 pages on vegetables in the 32-page catalog.

(9) Farmer Seed & Nursery Co.
Fairbault, MN 55021
84 pages, 8 x 10. Complete. Special attention to midget vegetables and early maturing varieties for the northern tier of states. Established 1888.

(10) Henry Field Seed & Nursery Co.
Shenandoah, IA 51602
128 pages, 8½ x 11. A complete catalog. Wide variety selection. Many hard-to-find items. Good tips for vegetable growers.

(11) DeGiorgi Co., Inc.
Council Bluffs, IA 51502
112 pages, 8½ x 11. Attention to the unusual. Established 1905. 66¢.

(12) Gurney Seed & Nursery Co.
1448 Page St.
Yankton, SD 57078
64 pages, 15 x 20. Emphasis on short-season North Country varieties. "Tried and True," "New and Novel," varieties of seed and nursery stock.

(13) Harris Seeds
3670 Buffalo Road
Rochester, NY 14624
84 pages, 8½ x 11. Vegetables 39 pages. The look of authority and so considered, especially in the Northeast.

(14) Charles C. Hart Seed Co.
Main & Hart Streets
Wethersfield, CT 06109
Vegetables, herbs and flowers, 24 pages

including All-America selections.

(15) H.G. Hastings Co.
Box 4274
434 Marietta Street, N.W.
Atlanta, GA 30302
68 pages, 8½ x 11. "91 years of service to the South." Complete southern garden guide.

(16) J.W. Jung Seed Co.
Station 8
Randolph, WI 53956
68 pages, 9 x 12. Everything for the garden. Over 1200 varieties of quality products at reasonable prices since 1907.

(17) Thompson & Morgan
P.O. Box 24
Somerdale, NJ 08083
160 pages, 5 x 7, accenting the unusual in vegetables, flowers and accessories. Also features Oriental vegetables and sprouting seeds.

(18) Earl May Seed & Nursery Co.
Shenandoah, IA 51603
80 pages, 8½ x 11. Wide choice of varieties. Features All-America Selections.

(19) Nichols Garden Nursery
1190 North Pacific Highway
Albany, OR 97321
88 pages, 8½ x 11. Written by an enthusiastic gardener and cook who has searched the world for the unusual and rare in vegetables and herbs.

(20) L.L. Olds Seed Co.
2901 Packers Avenue
Box 1069, Madison, WI 53701
80 pages, 8 x 10. Vegetables, 30 pages. Carefully written guide to varieties. All-America Selections.

(21) Geo. W. Park Seed Co., Inc.
Greenwood, SC 29647
124 pages, 8½ x 11¼. A guide to quality and variety in flowers and vegetables. Includes indoor gardening. The most frequently "borrowed" seed catalog.

(22) Reuter Seed Co., Inc.
320 N. Carrolton Avenue
New Orleans, LA 70119
32 pages, 8 x 10. Vegetables, 16 pages. "Serving the South Since 1881."

(23) Seedway
Hall, NY 14463
32 pages, 8½ x 11. Vegetables, 22 pages. Informative, straightforward presentation.

(25) J.L. Hudson
P.O. Box 1058
Redwood City, CA 94604
128 pages, 5½ x 9. Vegetables, 19 pages. General catalog $1. Accent on the unusual. Wide selection of herbs. Free Vegetable catalog.

(26) R.H. Shumway, Seedsman
628 Cedar Street
Rockford, IL 61101
84 pages 10 x 13. Complete. Founded 1870. Catalog has maintained some of the "good farming" 1870 look.

(27) Stokes Seeds
Box 548, 737 Main Street
Buffalo, NY 14240
150 pages, 5½ x 8½. Wide selections. Emphasis on short-season varieties. Canadian introductions.

(28) Otis S. Twilley Seed Co.
Salisbury, MD 21801
64 pages, 8½ x 11. Clear, helpful presentation with special attention to Experiment Station releases and disease-resistant varieties for varying climatic conditions.

(32) Geo. Tait & Sons, Inc.
900 Tidewater Drive,
Norfolk, VA 23504
Vegetables, 24 of 56 pages. Special varieties and planting information for eastern Virginia and North Carolina.

(33) Vesey's Seeds, Ltd.
York-P.E. Island
Canada
Features every vegetable variety. Local planting information.

(34) C.A. Cruickshank Ltd.
1015 Mount Pleasant Road,
Toronto, Canada M4P 2M1.
An 80-page "garden guild" general catalog.

(36) W. H. Perron & Co., Ltd.
515 Labelle Boulevard
Chomedey, P. Que., Canada H7V 2T3
Complete general catalog of 106 pages. $1.

(37) Porter & Son, Seedsmen
Stephenville, TX 76401
36 pages, 6 x 9. Vegetables, flowers, garden aids. Tomatoes of the region.

(38) T&T Seeds, Ltd.
111 Lombard Avenue
Winnipeg, Manitoba, Canada
60 pages, 6 x 8. Condensed general catalog for Midwest/North region. 35¢.

(39) Dominion Seed House
Georgetown, Ontario, Canada
80 pages on vegetables in 180-page general catalog. Will ship only to Canadian addresses.

(40) Laval Seeds, Ltd.
3505 Boul. St. Martin
Laval, Quebec, Canada
General catalog, 128 pages, 49 pages of vegetables. Printed in French only.

(41) MacFayden Seeds
P.O. Box 1600
30-9th Street
Brandon, MAN, Canada $1.

(42) Alberta Nurseries and Seeds Ltd.
Box 20 Bowden
Alberta, Canada TOM OKO
46 pages, 7 x 10. Vegetables, 14 pages. Special attention to hardiness and short-season. Items shipped prepaid. Free catalog.

(43) Agway, Inc.
Box 4933
Syracuse, NY 13221
56 pages, 8½ x 11. Thoughtfully prepared for northeastern states for their 700 retail outlets.

(44) Johnny's Selected Seeds
Albion, ME 04910
24 pages, 5½ x 8½. Traditional and heirloom American food and seed varieties plus Oriental seed varieties. 50¢.

(45) Kitazawa Seed Co.
356 W. Taylor Street
San Jose, CA 95110
One-sheet listing of Oriental vegetables, including bitter melon, Japanese pickling melon, gobo, and many more common Oriental vegetables.

(46) J.A. Demonchaux, Co.
837 North Kansas Avenue
Topeka, KS 66608
Gourmet garden seeds from France. 4-page list of vegetables and herbs.

(47) Grace's Gardens
60 Witton Road
Westport, CT 06880
World's most unusual seed catalog and clearing house for gigantic vegetables in the U.S. Also Oriental, Italian and Mexican vegetables. 25¢.

(48) Tsang and Ma International
P.O. Box 294
Belmont, CA 94002
Chinese vegetable seeds; each packet has planting and cooking information. Free catalog includes Oriental cookware and vegetable cookbook.

(50) Orol Ledden & Sons
Sewel, NY 08080
General catalog includes tools, supplies, and equipment. 34 pages, 8 x 10, 14 pages of vegetables.

(51) Archias Seed Store Corp.
P.O. Box 109
Sedalia, MO 65301
General catalog, 34 pages, 8 x 10. 19 pages of vegetables. Features Midwest tomato varieties.

(52) Horticultural Enterprises
P.O. Box 340082
Dallas, TX 75234
One-sheet folder featuring hot pepper varieties. Also includes seeds of tomatillo, jicama, cilantro.

(58) Clyde Robin Seed Co., Inc.
P.O. Box 3855
Castro Valley, CA 94566
100 pages, 4 x 9. Seed collection, woody plants and wildflowers. Also unusual herbs

and vegetables, 3 pages seed varieties of eucalyptus. $1.

(60) Vermont Bean Seed Co.
Way's Lane
Manchester, VT 05255
40 pages, 5 x 7. Specialists in all types of beans and peas. 25¢

(61) Kilgore Seed Company
1400 West First Street
Sanford, FL 32771
44 pages 8½ x 11 general catalog of vegetables, flowers, and many garden aids. Contains much useful information on Florida gardening.

(62) Herbst Brothers Seedmen, Inc.
1000 N. Main Street
Brewster, NY 10509
68 pages, 8½ x 11. General catalog of vegetables, flowers, and a wide variety of tree and shrub seeds. Large selection of garden aids.

(63) Epicure Seeds Ltd.
Avon, NY 14414
24 pages, 5½ x 8½. Specialists in imported European vegetable seeds. 25¢.

(64) Garden Gems Seeds
3902 State Street
Quincy, IL 62301
16 pages, 4 x 9. Flowers, ornamentals, herbs, 5 pages of selected vegetables.

(65) Redwood City Seed Company
P.O. Box 361
Redwood City, CA 94064
30 pages, 5½ x 8½. Miscellaneous catalog. 12 pages of rare and unusual and some imported vegetable seeds. 50¢.

(67) Midwest Seed Growers, Inc.
505 Walnut Street
Kansas City, MO 64106
23 pages, 8½ x 11. General catalog, 16 pages vegetables, 7 pages flowers.

(68) Exotica Seed Company
1742 Laurel Canyon Bl.
Hollywood, CA 90046
26 pages, 8½ x 11. Unusual plants from Ecuador, Mexico, and the Hawaiian Islands. Many exotic vegetable seeds. $1.

(69) Wyatt-Quarles Seed Company
P.O. Box 2131
Raleigh, NC 27602
32 pages, 8½ x 11. 12 pages vegetables, emphasis on southern varieties. Bulbs, annuals, live plants and roots. Flowers, seeds and herbs.

(70) John Brudy Exotics
P.O. Box 1348
Cocoa Beach, FL 32931
30 pages, 5 x 7. Catalog offers uncommon seeds of shrubs, trees and vines, including winged bean, jojoba and jicama. $1, refundable with $10 order.

(71) Le Jardin du Gourmet
P.O. Box 31
West Danville, VT 05873
8-page leaflet, 5 x 7. French varieties of vegetables, also beans. Herbs, plants and seeds.

(72) Wanigan Association
262 Salem Street
Lynnfield, MA 01940
22 pages, 5 x 7. Heirloom beans.

(73) Bauman's Pickle Room
P.O. Box 628
Spring Valley, CA 92077
Catalog contains 14 pages of vegetables, including tomatillo. Selection of books.

Handbook for Vegetable Gardeners

There will always be more questions than a book this size can answer. We hear questions like: "How can I preserve my harvest?", "Can I dry or freeze my extra zucchini?", or "How many cubic feet of organic matter do I need to mulch 27 square feet to a depth of 1 inch?" These questions usually come after a first try at growing vegetables at home.

Good Ideas from Good Gardeners

Good gardeners enjoy sharing information with others. We have received many good tips from many good gardeners. We have listed here the ones that seemed most helpful. After reading through the good ideas, check the charts on freezing, storing, and drying vegetables.

2" × 12"

½" wire mesh

Gopher Protection

"Two families of gophers practically ruined our garden last year. This year we have gopher-proofed one section by building a 12-inch deep raised bed and lining the bottom of the bed with ½-inch mesh wire."

Coffee Cans

"I am sure you have many ideas already. Hotkaps I am not crazy about as they can blow away if not carefully anchored and they sometimes get the plant too hot. We have had good luck with respect to late light frosts with 2-pound coffee cans, both tops and bottoms removed.

"We put them over our tomatoes, eggplants and peppers and leave them for a couple of weeks. Occasionally the tips of the leaves get burnt by the hot metal on contact but the damage is not permanent. We also use them on cabbage plants—not against frost but as protection against bugs. The advantage of the open-ended can is that you don't have to lift it every morning and replace it again each night. Of course it's not protection against a heavy frost, but one would not normally be putting out tender plants that early."

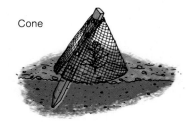

Cone

Cone Protection

"I have manufactured my own 'hot caps' by forming a cone out of plastic hardware cloth and a long pointed stick attached. It gives protection from frost and wind without danger of the heat build-up typical of hot caps. Furthermore, I can make the size fit the plant."

Bird Protection

"I made a cover of aluminum fly screen. I built portable frames 14 inches wide, 7 inches high, and 9 feet long for my peas and beans; and smaller ones 7 inches wide for my carrots, beets, and lettuce. I store them in the winter and bring them out for the first seed sowing. They're very simple.

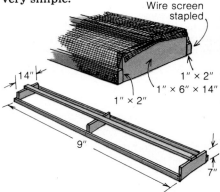

Wire screen stapled

14"

1" × 2"
1" × 6" × 14"
1" × 2"

9"

7"

"The frames should be ½, ⅓, or ¼ the length of a full row. Just staple the wire to frame. A heavy stapling gun is ideal and very fast. The 7-inch height is adequate. After the plants are 6 inches tall, birds don't do much damage. The first year we used 1-inch mesh chicken wire. It didn't stop the mice. Small birds squeezed through and couldn't get out again. But we have had no trouble after covering with the aluminum fly screen."

No Dirt, No Worms

"I have found a way to bring smiles to the gal in the kitchen. The vegetables I bring in are really clean. After kitchen inspection there's no dirt or worms in the sink.

"I have a large pail of water and a wire basket and I use either or both to suit the vegetable. Spray from the hose will take care of most of the dirt from root vegetables in the wire basket. Some may need to be washed in the pail. A couple of shakes and whirls in the wire basket gets rid of excess water."

Don't Worry About Clods

"Some gardeners work too hard spading and raking soil for a seed bed. I don't cheat on the depth of spading, but a few clods and small rocks don't bother me. When preparing seed bed for rows 2½ to 3 feet apart, I don't worry about clods between the rows.

"A clod-free strip can be developed quickly in very rough soil by hoeing up a ridge 4-6 inches high and then raking it down, pulling the clods off in the process. The row can be left slightly mounded if excess water is apt to be a problem."

Clothespins

"Spring-clip clothespins are useful when handling seed packets in the garden. Clip them on partly used packets—fold the top back—so the rest of the seeds won't spill out."

Handy Panels

"I hate to see transplants of anything wilt even temporarily. So when transplanting the thinnings from lettuce, beets and the like, I give them part shade for a few days. With small plants I stick a shingle at an angle to stop the hot sun. The best protection is with latticed panels. I make them 2 feet wide and 3 feet long."

Lath module

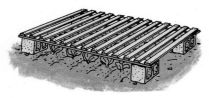

Ice Cream Scoop

"For setting out transplants of lettuce, tomatoes, cabbage, and the like, I prefer the ice cream scoop over the trowel. I can get the same depth in each planting hole fast and easy."

Little Red Wagon

"The little red wagon is a very practical gardening aid. In addition to simple transport from garage to garden and back again of small tools, fertilizers, containers, and such, we use it in early spring for indoor-outdoor movements of plants in pots in the transplant stage. We give them sunlight during the day and move them in the garage at night for frost protection."

Indoor Lettuce

An apartment dweller tells us: "I found that 'Ruby' leaf lettuce makes a beautiful house plant. I plant it in an 8-inch centerpiece bowl, water and fertilize it well and keep it near a window, and harvest a salad about once every three weeks all winter."

Nighttime Warmth

"We built a plastic A-frame (see p. 6) and found a way to keep it warm on cold nights. We put large plastic bleach bottles full of water inside the A-frame. The sun warms the water during the day. At night it slowly gives off heat and keeps the A-frame several degrees warmer than the outside air."

More Tomato Training

"We solved the problem of keeping tomatoes off the ground without staking and very little tying and pruning. I built two frames to place over twelve plants, set 26 inches apart in a 27-foot row. They're rugged and have lasted eight years. Here's a sketch of how they are made."

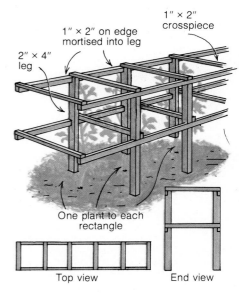

One plant to each rectangle

Top view   End view

Rotation

"Recommended by everyone, like motherhood, but more difficult to achieve, for the amateur. I have shown you my system of dividing garden into quarters and planting all that was in quarter #1 this year in quarter #2 next year, etc. The acetate overlay I use with my garden layout is not so much for record keeping as it is for planning. I find it invaluable."

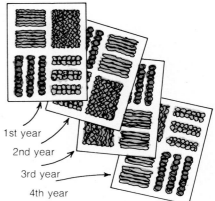

1st year
2nd year
3rd year
4th year

Hip-Pocket Tools

"For gardening in boxes and pots, the ordinary tools I could buy were too big and clumsy. I carry a 2-inch putty knife and a pair of scissors and that's all. I can harvest beet greens, prune tomato plants, and thin carrots and lettuce by snipping rather than pulling. The putty knife is a spade, cultivator, and hoe all rolled into one. It's perfect for blocking out transplants in a flat."

Parcel Post Storage

"We mounted the large parcel-post-sized mailbox on a 5-foot-high post alongside the main path in the garden. It gives us handy storage for insecticides, gadgets, twine, small tools, markers, and the like. It's weatherproof and high enough to keep things out of reach of small children. This year we trained a cucumber to climb up the post."

Old Rule

A seed company told us: "When seeds fail to germinate, some gardeners are sure that the fault must be in the seed. Actually it's almost impossible to buy seed that will not germinate. It is not uncommon to see 90 percent or more of the seeds produce excellent plants. Gardeners do, and should, sow more seeds than are usually needed for the final stand. This excess is good insurance against less-than-perfect soil conditions and other hazards. The old rule for sowing corn, beans, and peas still has some validity: One for the blackbird, one for the crow, one for the cutworm, and one to grow."

"What Variety Is That?"

"We try out new varieties in a rather cautious fashion. The trial plantings are small in comparison to the ones we have had success with. To make sure that we have the name of the variety at harvest time we keep a record on a garden map or on stakes.

"Seed packets on stakes quickly fade out in rain or sprinkling or get lost. We've tried several methods—from writing on stakes with a waterproof marker to protecting the seed packet. You can put the seed packet or written label between sheets of plastic (art supply stores carry plastic sheets suitable for this purpose.) Or cover the packet with mat-surface, scotch-type tape. Or use lacquer in a spray can. Fasten the packet or label to the stake with double-sided tape or white glue."

Plastic Netting

"For the first time, we used plastic bird netting to protect newly planted corn, peas, and beans. We completely foiled such nuisance birds as brown thrashers, and for the first time in many years replanting was unnecessary."

A strawberry container offers another kind of harvest when planted with Japanese cucumber, herbs and pink banana squash.

One way to keep a melon from getting sunburned. A simply constructed "A" frame, pictured, can be moved from place to place.

"Soaker" hose is very effective for watering young plants. It's portable and provides slow irrigation in long rows.

A coffee can waterer with holes punched in the sides. No need to trench, no erosion, less weeds. It drains slowly and confines water to the roots.

Black plastic stops weeds, acts as a mulch and slows evaporation. It increases yields of many warm weather crops, especially melons.

A bushel basket is ideal for a "Potpourri Salad" basket. This one is planted with radish, red lettuce, endive and parsley.

Vegetable	Suitability for freezing	Comments
Asparagus	excellent	
Beans, green	good	Tendercrop and closely related varieties and Blue Lake varieties either bush or pole are preferred because of their good flavor. Also, Blue Lake has a desirable thick flesh.
Beans, wax	good	
Lime Beans, green	excellent	Fordhook types preferred.
Beets	fair	Better canned; select small roots only for freezing.
Broccoli	excellent	
Cabbage	not recommended	Preserve as sauerkraut.
Carrots	fair	Select tender roots only. Can be diced and frozen with peas.
Cauliflower	excellent	Also suitable for pickling.
Celery	not recommended	Except in "soup packages."
Chinese Cabbage	not recommended	Use in sauerkraut.
Cucumbers	not recommended	Preserve by picking (see notes below).
Eggplant	fair	Significant quality loss; suitable for casserole dishes.
Endive	not recommended	
Kale	good	Select young leaves only.
Kohlrabi	fair	Significant quality loss, picks up *high* flavor.
Lettuce	not recommended	
Muskmelon	fair	Firm fleshed varieties are preferred; freeze small pieces; use within 3 months.
Mustard	good	Select tender leaves and remove stems.
Onions	fair	Freeze chopped, mature onions; significant quality loss; use in 3 months.
Parsley	not recommended	Can be dried.
Parsnips	fair	Significant quality loss.
Peas	excellent	Frosty and Perfected Freezer 60 preferred. All large wrinkled seeded varieties are suitable and so are edible podded varieties.
Peppers	fair	Significant quality loss; better if frozen chopped; use in 3 months.
Pop Corn	not recommended	
Potatoes	not recommended	Store fresh at 40-50° F.
Pumpkins	not recommended	
Radishes	not recommended	
Rhubarb	excellent	Varieties with red stalks like Canada Red, Valentine, and Ruby preferred. Pull stalks soon after they reach full size.
Swiss Chard	good	Select only tender leaves; remove midrib or stems.
Spinach	excellent	Savoy varieties are often preferred.
Summer squash	fair	Significant quality loss.
Winter squash	good	Be sure squash is fully mature (hard rind); freeze cooked pieces or mash.
Sweet corn	good to excellent	Jubilee, Seneca Chief, Golden Cross, and Silver Queen preferred; corn on cob frozen without blanching should be eaten in 6-8 weeks.
Tomatoes	fair	Better canned; freeze only juice or cooked tomatoes.
Turnips & Rutabagas	fair	Significant quality loss.
Watermelon	fair	Freeze only as pieces; use within 3 months.

Notes:

The term "significant quality loss" means the product after being frozen is quite inferior to the fresh product.

For cucumber pickles, use pickling varieties if many pickles are to be made, though young slicing cucumbers are suitable for quick-method dills.

In canning, the variety is seldom of major consideration for quality in home canning. Vegetables harvested at peak of quality and processed promptly usually will provide a high quality product regardless of variety. Tomato varieties should be chosen with the family preferences for mild or acid flavors in mind.

Chart by Ruth Klippstein and P. A. Minges, Home Garden Dept. of Vegetable Crops, Cornell University.

FREEZING VEGETABLES

Today, freezing is generally thought to be the best method for preserving flavor and color in vegetables. When you freeze them, choose only top quality to start with; don't bother to freeze those that are old or of questionable quality. Freeze vegetables as soon after picking as possible. Always try to pick them during the cool part of the day.

Also keep in mind that each vegetable has a specific method of preparation and blanching before it can be frozen. For more information see Ortho's *12 Month Harvest*.

STORING VEGETABLES

Under the proper conditions, many vegetables can be stored for quite a long time. A cool basement can be an excellent place to store vegetables —so can the kitchen refrigerator. Some people also build special underground storage pits.

A Michigan gardener stores root crops to be harvested fresh from the garden throughout the winter months. He writes: "Letting mother nature store vegetables year round is no new trick. The pioneers had to depend on this method in order to have a winter and spring supply of food. Come late October, when winter sets in in Michigan, we have an ample supply of carrots, leeks, onions, kale, Jerusalem artichokes, and parsnips left in the ground to see us through to next years first harvest. The vegetables don't grow during the winter months but with a few inches of mulch to keep the ground from freezing, the crop will be kept fresh and can be harvested as needed."

Listed at right are the ideal storage conditions for many of the vegetables described in this book.

For more information consult the USDA Handbook #66 or your local extension agent.

Sid Harkema gathering parsnips from straw mulch in the winter.

DRYING VEGETABLES

A surprisingly large number of vegetables can be dried easily. Dried vegetables are great in soups and a good addition to camping trips.

Home driers are available in many garden stores, mail-order seed catalogs, and department stores. For best results, pick the vegetables in their prime, blanch them if necessary, and dry them immediately. See Ortho's *12 Month Harvest* for more information on drying vegetables.

Storage Recommendations and the Approximate Lengths of the Storage Period

Vegetable	Temperature (F°)	Humidity (%)	Approximate Length of Storage Period
Cold, Moist Storage			
Asparagus	32-35	85-90	2-3 weeks
Beets, topped	32	95	3-5 months
Broccoli	32-35	90-95	10-14 days
Brussels sprouts	32-35	90-95	3-5 weeks
Cabbage, late	32	90-95	3-4 months
Cabbage, Chinese	32	90-95	1-2 months
Carrots, mature and topped	32-35	90-95	4-5 months
Cauliflower	32-35	85-90	2-4 weeks
Celeriac	32	90-95	3-4 months
Celery	32-35	90-95	2-3 months
Collards	32-35	90-95	10-14 days
Corn, sweet	32-35	85-90	4-8 days
Endive, Escarole	32	90-95	2-3 weeks
Greens, leafy	32	90-95	10-14 days
Horseradish	30-33	90-95	10-12 months
Kale	32	90-95	10-14 days
Kohlrabi	32	90-95	2-4 weeks
Leeks, green	32	90-95	1-3 months
Lettuce	32-35	90-95	2-3 weeks
Onions, green	32-35	90-95	3-4 weeks
Parsnips	32-35	90-95	2-6 months
Peas	32-35	85-90	1-3 weeks
Potatoes, late crop	35-40	85-90	4-9 months
Radishes	32-35	90-95	3-4 weeks
Rhubarb	32-35	90-95	2-4 weeks
Rutabagas	32-35	90-95	2-4 months
Spinach	32-35	90-95	10-14 days
Turnips	32	90-95	4-5 months
Cool, Moist Storage			
Beans, snap	40-45	90-95	7-10 days
Beans, lima	32-40	90	1-2 weeks
Cantaloupe	40	85-90	15 days
Cucumbers	40-50	85-90	10-14 days
Eggplant	40-50	85-90	1 week
Okra	45	90-95	7-10 days
Peppers, sweet	40-50	85-90	2-3 weeks
Potatoes, early	50	90	1-3 weeks
Potatoes, late	40	90	4-9 months
Squash, summer	40-50	90-95	5-14 days
Tomatoes, ripe	40-50	85-90	4-7 days
Tomatoes, unripe	60-70	85-90	1-3 weeks
Watermelon	40-50	80-85	2-3 weeks
Cool, Dry Storage			
Beans, dry	32-40	40	Over 1 year
Garlic, dry	32	65-70	6-7 months
Onions, dry	32	65-70	1-8 months
Peas, dry	32-40	40	Over 1 year
Peppers, dry chili	32-50	60-70	6 months
Shallots, dry	32	60-70	6-7 months
Warm, Dry Storage			
Pumpkins	55-65	40-70	2-4 months
Squash, winter	55-65	40-70	3-6 months
Sweet potato	55-60	70-85	4-6 months
Tomatoes, unripe	55-70	85-90	1-3 weeks

SOURCE: Adapted from Wright, R.C., D.H. Rose and T.M. Whiteman, 1954. "The commercial storage of fruits, vegetables and florist and nursery stock." *USDA Handbook* No. 66.

Proof-of-Purchase

0-917102-90-8